OECD Studies on Water

AF695192

Diffuse Pollution, Degraded Waters

EMERGING POLICY SOLUTIONS

OECD
BETTER POLICIES FOR BETTER LIVES

This work is published under the responsibility of the Secretary-General of the OECD. The opinions expressed and arguments employed herein do not necessarily reflect the official views of OECD member countries.

This document and any map included herein are without prejudice to the status of or sovereignty over any territory, to the delimitation of international frontiers and boundaries and to the name of any territory, city or area.

Please cite this publication as:
OECD (2017), *Diffuse Pollution, Degraded Waters: Emerging Policy Solutions*, OECD Studies on Water, OECD Publishing, Paris.
http://dx.doi.org/10.1787/9789264269064-en

ISBN 978-92-64-26905-7 (print)
ISBN 978-92-64-26906-4 (PDF)

Series: OECD Studies on Water
ISSN 2224-5073 (print)
ISSN 2224-5081 (online)

Photo credits: Cover © Wolfilser / Shutterstock.com

Corrigenda to OECD publications may be found on line at: *www.oecd.org/about/publishing/corrigenda.htm.*

© OECD 2017

You can copy, download or print OECD content for your own use, and you can include excerpts from OECD publications, databases and multimedia products in your own documents, presentations, blogs, websites and teaching materials, provided that suitable acknowledgement of OECD as source and copyright owner is given. All requests for public or commercial use and translation rights should be submitted to *rights@oecd.org*. Requests for permission to photocopy portions of this material for public or commercial use shall be addressed directly to the Copyright Clearance Center (CCC) at *info@copyright.com* or the Centre français d'exploitation du droit de copie (CFC) at *contact@cfcopies.com*.

Foreword

Water of adequate quality is an increasingly scarce resource. Substantial investments in wastewater treatment plants and progress in controlling point sources of pollution have contributed to significant improvements in water quality in recent decades. But a focus on point source pollution as a means of improving water quality is reaching its limits. Water pollution from unregulated diffuse sources of pollution from both urban and rural areas continues to rise. Unless attention is turned to these sources, further deterioration of water quality and freshwater ecosystems can be expected as human populations grow, industrial and agricultural production intensifies, and climate change causes significant alteration to the hydrological cycle.

Unlike point source pollution, which enters a water body at a specific site such as a pipe discharge, diffuse pollution occurs when pollutants from a variety of activities runoff, leach or deposit into surface and groundwater bodies. The most prevalent water quality challenge globally is eutrophication. This is characterised by oxygen depletion and algal blooms leading to significant loss of aquatic biodiversity. The primary cause can be traced to excess nutrients from agricultural runoff.

Reducing the costs of diffuse pollution requires much greater attention from policymakers. The cost of current water pollution from diffuse sources exceeds billions of dollars each year in OECD countries. Economic costs include: degradation of ecosystem services; health-related costs; impacts on economic activities such as agriculture, industrial production and tourism; increased water treatment costs; and reduced property values, among others. The scale of these costs means that seeking increasingly marginal reductions in point source pollution is no longer the most cost-effective approach to improving water quality in many OECD countries.

The relative lack of progress with reducing diffuse pollution reflects the complexities of controlling multiple pollutants from multiple sources, their high spatial and temporal variability, associated transactions costs, and limited political acceptability of regulatory measures.

This report, "Diffuse Pollution, Degraded Waters: Emerging Policy Solutions" takes a major step forward in providing policy guidance on better managing water quality risks and navigating the challenges of diffuse pollution. It reveals that many current policy responses to address diffuse pollution do not reflect some of the basic principles of water quality policy, such as the Polluter Pays Principle, and largely rely on voluntary participation and compliance measures.

The report highlights emerging policy solutions, such as a natural capital based approach to allocating diffuse pollution limits to individual property owners, water quality trading, pollution charges, collaborative governance, and outcome-oriented contributions to policy design. It provides a risk-based framework for intervening and policy principles to guide policymakers and stakeholders through the myriad decisions required to establish new or alter existing water quality management regimes. The report stresses that economic instruments, such as pollution charges or tradable entitlements, are an under-utilised means of increasing the cost effectiveness of pollution control strategies while simultaneously promoting innovation.

Limiting diffuse water pollution within acceptable boundaries is essential. While water quality goals are obviously at the core of a policy response, many other sectoral policy frameworks need to be aligned if efforts to reduce the costs of diffuse pollution are to be fruitful. I am confident that policymakers can find both inspiration and pragmatic support in this report.

Improving water quality is a critical element of the 2030 Sustainable Development Goals, fulfilling an essential role in reducing poverty and disease and promoting sustainable growth. It is also a key element of the OECD's recently adopted "Council Recommendation on Water". These significant commitments frame the water agenda which, at its core, calls for the integrated, sustainable and equitable management of water.

Simon Upton,
OECD Environment Director

Acknowledgements

This report is an output of the OECD Environment Directorate, under the leadership of Director Simon Upton and the subsidiary Climate, Biodiversity and Water Division led by Simon Buckle. The project co-ordinator and author of the report is Hannah Leckie. The author is thankful in particular to Xavier Leflaive (Water Team Leader, OECD Environment Directorate), for his substantive inputs and guidance throughout the development of the report.

The author wishes to acknowledge the valuable contribution of delegates from the Working Party on Biodiversity, Water and Ecosystems and the Joint Working Party on Agriculture and Environment in preparation of this report. The financial and in-kind contributions of the governments of Korea and the Netherlands helped to make this work possible and are greatly appreciated.

The author expresses gratitude to those who contributed case studies for the project, including Ian Barker (Water Policy International Ltd), Alistair Boxall (The University of York), Nick Cartwright (Environment Agency, United Kingdom), Paul Chapman (London Borough of Lewisham), Maree Clark and Sean Hodges (Horizons Regional Council, New Zealand), the European Commission, Dmitry Frank-Kamenetsky (Baltic Marine Environment Protection Commission, Finland), Jim Gebhardt (Environmental Protection Agency, United States), Nick Haigh (Defra, United Kingdom), Tony Harrington (Welsh Water), Maude Jolly and Emmanuel Steinmann (Ministry of Ecology, France), Alec Mackay (AgResearch, New Zealand), Brendan McHugh (Marine Institute, Ireland), Nic Newman (Environment Canterbury, New Zealand), Jihyung Park and Sang-Cheol Park (Ministry of Environment, Korea), Fiona Regan and Lisa Jones (Dublin City University), Ellen van Lindert (Ministry of Infrastructure and Environment, Netherlands), Annemarie van Wezel (KWR Watercycle Research Institute and Copernicus Institute/Utrecht University, Netherlands), Sara Walker (World Resources Institute), Adi Yefet (Israel New Tech), Sarantuyaa Zandaryaa (UNESCO), and Alon Zask (Ministry of Environmental Protection, Israel). Some of these case studies will be relevant to future policy work on contaminants of emerging concern in surface waters.

The project benefitted from insightful discussions and the wisdom of participants at the *OECD Workshop on Innovative Policy Instruments for Water Quality Management* supported by the Netherlands. The workshop, held in March 2016 in The Hague, helped to develop a policy framework to manage diffuse pollution. The support of Ellen Van Lindert (Ministry of Infrastructure and Environment, Netherlands) was critical to the success of the workshop.

The author is also grateful to colleagues and experts who provided valuable comments, in particular Guillaume Gruère, Kathleen Dominque, Michael Mullen and Delphine Clavreul (OECD Secretariat), Dave Tickner (WWF) and Tony Harrington (Welsh Water). Editorial guidance from Janine Treves, Beth Del Bourgo and Stéphanie Simonin-Edwards, and administrative support from Sama Al Taher Cucci and Angèle N'Zinga are gratefully acknowledged.

Table of contents

Acronyms and Abbreviations ... 9

Executive summary ... 11

Chapter 1: The Water Quality Challenge ... 13
- **Key messages** ... 14
- An introduction to water quality and its impact on the environment and society ... 15
- A typology for water pollution: sources, types and pathways ... 16
- Ongoing challenges of diffuse pollution sources and eutrophication in OECD countries ... 20
- Water quality and climate change ... 23
- Water quality challenges for cities of OECD countries ... 25
- Notes ... 29
- References ... 30

Annex 1.A1: An overview of the main water pollutants in OECD countries ... 35

Chapter 2: Economic costs and policy approaches to control diffuse source water pollution ... 47
- **Key messages** ... 48
- The economic case for water quality management ... 49
- Water pollution control mechanisms in OECD countries ... 59
- The need for policy coherence ... 62
- Note ... 65
- References ... 65

Chapter 3: Emerging policy instruments for the control of diffuse source water pollution ... 69
- **Key messages** ... 70
- Introduction ... 71
- Regulatory approaches ... 72
- Economic instruments ... 81
- A collective management approach to water quality management ... 94
- Notes ... 98
- References ... 99

Chapter 4: A policy framework for diffuse source water pollution management ... 101
- **Key messages** ... 102
- A policy framework for diffuse source water pollution management ... 104
- A role for central government ... 116
- Note ... 118
- References ... 118

Tables
- 1.1. Water pollutants and their typical sources ... 19
- 1.2. Effects of climate change on water quality ... 25
- 2.1. Impacts of water pollution: Economic, social and environmental ... 49
- 2.2. Use values and non-use values of water ecosystems in relation to its quality ... 52
- 2.3. Estimated annual national costs of water pollution: A selection from OECD countries ... 54
- 2.4. Range of cost and effectiveness of management practices to mitigate phosphorus losses from New Zealand dairy farms ... 55
- 2.5. Asset values of UK freshwater ecosystems (wetlands and open channels) 2008-12 ... 57
- 2.6. Regulatory, economic, and voluntary pollution control mechanisms, directed at source and end-of-pipe ... 58

2.7. Advantages and challenges with the Polluter Pays and Beneficiary Pays Principles for the control of diffuse pollution ... 61
2.8. Examples of potential trade-offs and co-benefits between climate change mitigation/adaptation measures and water quality ... 64
3.1. Allocation approaches for nutrient discharge allowances ... 78
3.2. Cumulative nitrogen leaching maximum by Land Use Capability Class, Manawatu-Wanganui Region, New Zealand ... 80
3.3. Examples of features of pollution charges in selected OECD countries ... 82
3.4. Clean Water and Drinking Water State Revolving Funds: Funds Available for Projects 2010-2015 ... 91
4.1. A policy framework to manage diffuse water pollution ... 104
4.2. The Polluter Pays Principle for diffuse water pollution: Barriers and solutions ... 109
4.3. Examples of water quality policies and trade-offs and co-benefits to other sectors ... 111
4.4. Policy instruments to address diffuse water pollution and protect freshwater ecosystems ... 113
4.5. Some requisites for water pollution charges and taxes ... 114
4.6. Some requisites for diffuse pollution allocation and water quality markets ... 114

Figures

1.1. Global distribution of water pollution hazard, 2000 ... 15
1.2. A typology for water pollution ... 17
1.3. Percentage share of agriculture in total emissions of nitrates and phosphorus in surface water, OECD countries, 2009 or latest available year ... 21
A1. The principle processes causing groundwater salinisation ... 38
2.1. Response time delays in water pollution abatement (societal and ecological) ... 50
2.2. Change in median dissolved reactive phosphorus concentration and profitability 5 years after the consecutive implementation of three mitigation strategies in a catchment draining a dairy farm, Otago, New Zealand ... 56
3.1. Water quality improvements under the Total Maximum Daily Load Management System, Korea ... 78
3.2. Cost comparison of options to reduce a nitrogen loading, Chesapeake Bay Watershed, United States ... 85
3.3. The stakeholder engagement staircase ... 95
4.1. A Risk-Based Approach to water quality management ... 105

Follow OECD Publications on:

http://twitter.com/OECD_Pubs

http://www.facebook.com/OECDPublications

http://www.linkedin.com/groups/OECD-Publications-4645871

http://www.youtube.com/oecdilibrary

OECD Alerts http://www.oecd.org/oecddirect/

Acronyms and abbreviations

AMD	Acid mine drainage
BOD	Biochemical oxygen demand
CECs	Contaminants of emerging concern
CSO	Combined sewer overflow
CWMS	Canterbury Water Management Strategy
CWSRF	Clean Water State Revolving Fund
DDT	Dichlorodiphenyltrichloroethane
DOC	Dissolved organic carbon
DRP	Dissolved reactive phosphorus
DWSRF	Drinking water State revolving fund
EPA	United States Environmental Protection Agency
EU	European Union
IPCC	Intergovernmental Panel on Climate Change
IUCN	International Union for Conservation of Nature
LRI	Land Resource Inventory
LUC	Land Use Capability
NCA	Natural capital accounting
NDA	Nutrient discharge allowance
NO_x	Nitrogen oxides
NH_3	Ammonia
N_2O	Nitrous oxide
NVZ	Nitrate vulnerable zone
OECD	Organisation for Economic Co-operation and Development
PCBs	Polychlorinated biphenyls
PES	Payment for ecosystems services
POPs	Persistent organic pollutants
PPP	Public-private partnership
R&D	Research and development
SRF	State Revolving Fund
TMDL	Total maximum daily load
TTT	Thames Tideway Tunnel
UK	United Kingdom
UNESCO	United Nations Educational, Scientific and Cultural Organization
WFD	Water Framework Directive
WHO	World Health Organisation
WWTP	Wastewater treatment plant

Executive summary

Decades of regulation and large investments to reduce point source water pollution have brought substantial gains for the economy, human health, environment and social values. But water quality challenges endure in OECD countries as a result of under-regulated diffuse sources of pollution. Eutrophication, a form of water pollution due mainly to agricultural runoff of excess nutrients, is the most prevalent challenge globally.

Controlling diffuse pollution is a complex task. Diffuse pollution is comprised of multiple pollutants from diverse sources – mainly agricultural and urban runoff – and varies spatially and over time. Regulating such pollution generally entails high transaction costs and often meets with political resistance. Lax enforcement of the regulatory measures that are in place weakens their impact. Climate change puts further pressure on water quality, exacerbating existing challenges due to altered precipitation, flow and thermal regimes, and sea level rise.

The cost of current water pollution from diffuse sources exceeds billions of dollars each year in OECD countries. Water pollution has lasting negative impacts on human health, water security, economic productivity, freshwater ecosystem services (including their ability to process pollutants) and social values. Polluted water decreases benefits from swimming, fishing and other recreational uses of water bodies and drags down property values of nearby real estate.

To date, policies in OECD to control diffuse pollution have typically fallen short of addressing the challenge and fail to fully reflect the Polluter Pays Principle. A heavy reliance on voluntary measures is pervasive. Lag times between pollution that degrades the resource and control measures that improve it exacerbate the challenges of managing diffuse water pollution, particularly in terms of who benefits from quality improvements and who pays for them. There is an urgent need to find cost-effective policies and measures to fund water quality improvements.

This report, *Diffuse Pollution, Degraded Waters: Emerging Policy Solutions*, addresses the water quality challenges facing OECD countries from diffuse water pollution. It examines the trends, drivers and impacts of water pollution and analyses a range of policy instruments to control diffuse pollution, illustrated by several case studies of innovative approaches. The report presents a risk-based framework that can assist policy makers and stakeholders to establish new, or strengthen existing, water quality management regimes. The key elements to successful reform of water quality management policies are:

- **Political ambition.** Completely eliminating water pollution risks is often neither technically possible nor cost-effective. Setting the appropriate level of ambition is ultimately a political decision and these decisions should be guided by an assessment the risks (environmental, economic and social), the cost of any resulting improvements in water quality, and society's level of acceptable risk. A lack of full scientific certainty should not be used as a reason for postponing action. Making links with higher level policy issues – such as public health, food security, energy production and tourism – can provide the stimulus and strengthen the case for political action. Citizen science, and advances in sensor technology, earth observations and water quality and economic modelling provide new data to inform priorities for action.

- **Policy principles**. A set of six principles can guide the design and implementation of policies to control diffuse water pollution:
  - The Principle of Pollution Prevention underscores the fact that the prevention of diffuse pollution is often more cost effective than treatment and restoration options.
  - The Principle of Treatment at Source encourages treatment at the earliest stage possible, which is generally more effective and less costly than waiting until pollution is widely dispersed.
  - The Polluter Pays Principle makes it costly for those activities that generate diffuse pollution and provides an economic incentive for reducing the pollution.
  - The Beneficiary Pays Principle allows sharing of the financial burden with those who benefit from water quality improvements. Requiring minimum regulatory standards to reduce pollution be met before payments are made is necessary to ensure additionality and avoid rewarding polluters.
  - Equity among different groups and across generations should be considered in the allocation of pollution rights and the costs and benefits of abatement.
  - Policy coherence across sectors is essential in ensuring that initiatives taken by different agencies (e.g. water, agriculture, urban planning and climate) do not have inadvertent negative impacts on water quality and can capitalise on potential co-benefits from water quality interventions.
- **Mix of policy instruments**. Regulatory, economic and voluntary policy instruments are all part of the toolkit that is needed to manage multiple sources of diffuse water pollution. The report highlights that economic instruments, such as pollution taxes, charges and water quality trading, could be strengthened and used more extensively to increase the cost effectiveness of pollution control and promote innovation. Advances in computer modelling offers an opportunity to design policy instruments directly proportional to the amount of estimated pollution generated or reduced from individual properties within a catchment. An allocation approach that captures the inherent differences in the underlying natural capital stocks of soils offers an approach to account for the full economic potential of natural resources.

Central government has a critical role to play in the transition to more effective management of the risks to water quality from diffuse pollution. This report lays out the recommended steps to meet this challenge. These include: *i)* providing overarching national policy guidance and minimum standards; *ii)* creating a robust institutional framework ; *iii)* engaging stakeholders to manage perceived and actual risks; *iv)* signalling policy changes and highlighting options for implementation; and *v)* implementing robust policies that minimise the cost of water quality management and promote innovation.

Diffuse Pollution, Degraded Waters
Emerging Policy Solutions
© OECD 2017

Chapter 1

The water quality challenge

This chapter takes stock of recent information and data on challenges related to water quality in OECD countries. It zooms in on the water quality issues facing OECD cities, the effects of water quantity and climate change on water quality, and the ongoing challenge of managing diffuse pollution, in particular, nutrient loading.

Key messages

Water quality continues to deteriorate despite improvements in the control of industrial point source pollution and wastewater treatment. Ongoing water quality problems in OECD countries are characterised by a number of pollutants, none more so than nutrient pollution, primarily from agricultural sources, which leads to eutrophication and harmful algal blooms. As a consequence, the relative importance of diffuse pollution loads is increasing in OECD countries, and increasing treatment and regulation of point source pollution is no longer necessarily the most cost-effective approach to improving water quality. However, maintaining these processes to manage point source pollution is essential and must not be abandoned.

OECD cities face distinct challenges, given that the negative impacts of poor water quality largely fall on cities (e.g. increased water treatment costs, health service costs), as does the value of assets at risk (e.g. corrosion and premature ageing of infrastructure and reduced property values from contaminated water), and the costs of treating pollution (e.g. wastewater and stormwater) before discharging to the environment. Diffuse pollution from stormwater runoff and combined sewer overflows is an ongoing challenge for cities. Climate change will exacerbate existing water quality challenges, due to altered precipitation, flow and thermal regimes, and sea level rise, which will mean water authorities and water and sanitation utilities will be confronted by further economic and operational challenges.

Freshwater of a high quality is also valued for environmental uses, such as the provision of fish habitat and ecosystem health. However, freshwater ecosystems are under immense pressure as a result of a legacy of industrial pollution and alteration of the natural morphology of water bodies, continuing pollution from diffuse sources (agricultural and urban), and an ever-evolving number of emerging pollutants in wastewater. This pollution, coupled with the effects of hypoxia, algal blooms, the introduction of invasive alien species and climate change, are having a devastating impact on freshwater biodiversity. Policy responses to these complex water quality challenges are required to protect freshwater ecosystems and the services they provide.

An introduction to water quality and its impact on the environment and society

Good water quality is essential for human well-being, for use in agriculture, aquaculture, and industry, and to support freshwater ecosystems and the services they provide. What qualifies as "good" water quality depends on the purpose of use and the value society holds for water quality.

Water pollution is defined as anthropogenic contamination[1] of water bodies (e.g. rivers, lakes, groundwater, estuaries and oceans) from the discharge, directly or indirectly, of a substance that changes the functioning of the system (Hanley et al., 2013). Pollution alters the composition and characteristics of a water body and its level of water quality. For example, the discharge of organic waste from sewers to rivers accelerates biological processes, and in the process uses up oxygen which can cause loss of aquatic life. Nutrients from fertilisers and livestock from the agriculture sector can lead to eutrophication of rivers and lakes, and can result in toxic algal blooms and changes in freshwater fauna and flora communities. Furthermore, poor water quality reduces the quantity of useable water and therefore exacerbates the problem of water scarcity.

Figure 1.1 illustrates the global distribution of pollution, which includes the effects of nutrient and pesticide loading, mercury deposition, salinisation, acidification, and sediment and organic loading (Sadoff et al., 2015; Vörösmarty et al., 2010). Pollution "hotspots"[2] are identified in most regions of the world, including OECD countries.

Figure 1.1. **Global distribution of water pollution hazard, 2000**

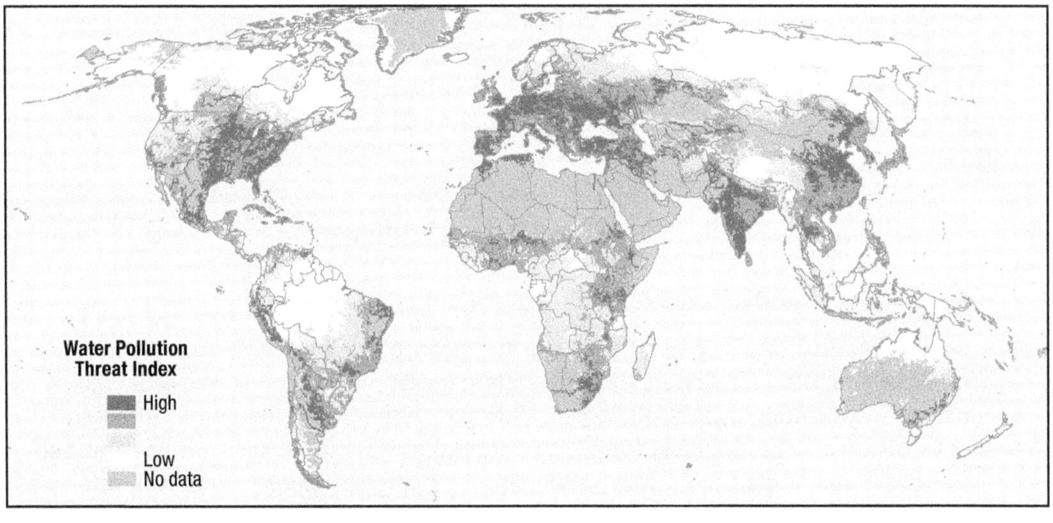

Source: Sadoff et al. (2015); based on data from Vörösmarty et al. (2010).

Population growth, coupled with climate change[3], are thought to have the greatest effect on water quality (Allan, et al., 2013), placing increasing pressure on the ability of finite water bodies to process wastewater, nutrients and contaminants before they lose their life-supporting function. So much so, that at least half the world's population suffers from polluted water (Jones, 2009). And the situation is set to worsen. Under even the most optimistic economic growth and climate change scenarios, a global and rapid increase in nitrogen (35-46%), phosphorus (15-24%) and biochemical oxygen demand (BOD[4]) (9-11%) is projected to 2050 (IFPRI and Veolia, 2015). Increases are projected in all regions of the world, but will be felt the greatest in upper-middle and lower-middle income countries, particularly Asia. This will, in turn, increase risks to human health, economic development and ecosystems.

Pollution, over-exploitation and alteration of water bodies as a result of human activities have led to the extinction, or risk thereof, of 10 000 to 20 000 freshwater species (Strayer and Dudgeon, 2010; Vörösmarty et al., 2010) - an 81% reduction in freshwater biodiversity between 1970 and 2012 (WWF, 2016). Of further concern, wetlands, which are biodiversity hotspots that deliver a wide range of ecosystem services including water purification, have declined by 64% globally since 1900 (Ramsar, 2015). Polluted freshwater also has an impact on coastal and ocean waters, for example the formation of eutrophic and hypoxic zones (also known as "dead zones") in the oceans.

Improving water quality is consistently ranked as a top environmental concern in public opinion surveys across most OECD countries (OECD, 2012a). For example, in the United States, an annual national public opinion survey from 1989 to 2014 consistently ranked water pollution as one of the top environmental concerns from a list including climate change, loss of rain forest, extinction of plant and animal species, and air pollution (Gallup Poll, 2014). A similar survey in the European Union in 2012 showed comparable results to those of the United States, with 84% of respondents listing chemical pollution as the greatest threat to a country's water environment, ahead of climate change, changes to the water ecosystem, floods, water scarcity, and other water-related threats (European Commission, 2012a). Challenges to water quality are the primary environmental concern for New Zealanders, ahead of air quality, terrestrial biodiversity, coastal waters and soils, with public attention increasingly focusing on the impact of agricultural runoff (Hughey et al., 2013).

Over recent decades, policy actions and major investment in OECD countries have helped to reduce point source pollution from urban centres, industry and wastewater treatment plants, with substantial gains for the economy, human health, environment and social values linked to water (OECD, 2012b). However, despite these improvements, diffuse pollution loads from agricultural and urban sources, combined sewer-overflows, and emerging contaminants in human and animal wastewater are continuing challenges in OECD countries (OECD, 2012b).

A typology for water pollution: sources, types and pathways

Characteristics and determinants of water quality

The quality of a water body is the function of its physical, biological and chemical characteristics. Physical characteristics relate to temperature, colour, taste, odour, turbidity and salinity, among others. Biological characteristics relate to living organisms, such as bacteria, zooplankton, algae, fungi, invertebrates, worms, aquatic plants and fish, among others. Chemical characteristics relate to pH, biological oxygen demand, and substances that dissolve in water, such as total dissolved solids, dissolved oxygen, nitrates, phosphates and other minerals.

Contaminants from naturally occurring events and human activities alter these characteristics, with a corresponding change in the composition of the waterbody and its level of water quality (Joyce and Convery, 2009). The nature of these alterations is not always linear, and can depend on a combination of variables related to the characteristics, volume and concentration of the pollutants (individually and in combination), the characteristics of the receiving water body, distance to the polluting source, the stochastic environmental conditions and timing (as outlined in Figure 1.2). Pressures from a range of policies and developments can affect water quality, such as water allocation, flood management, urban development, alterations to the natural morphology of water bodies, land and soil management practices, and climate change.

Figure 1.2. **A typology for water pollution**

Pollutant characteristics	Source type and pathways	Receiving body type and characteristics	Environmental conditions
• Toxicity • Concentration • Volume of discharge • Life span • Fate and transport • Ability to treat with current technologies • Chemical reactions (adsorption, dissolution, precipitation, decay) • Stock or Flow pollutant • Ambient or exogenous • Continuous or intermittent	**Type** • Point source • Diffuse source • Historic pollution **Pathways** • Pipe discharges • Surface runoff • Subsurface flow • Leaching • Dry and wet deposition (of atmospheric pollutants) • Re-suspension of contaminated sediment	**Type** • River • Lake • Groundwater • Wetland • Estuary/sea/ocean **Characteristics** • Physical, biological and chemical properties (ecosystem health) • Biological processes (processing pollutants, plant uptake, nutrient cycling, adsorption, mineralisation) • Natural contaminant background levels • Geographical features (morphology, topography, mountain-fed, glacier-fed, lowland, upstream or downstream) • River channel type (straight, meandering, braided) • Perennial or ephemeral • Surface-groundwater interactions • Water body modifications (e.g. dams, canals, dredging) • Lake stratification and mixing • Flow rate and residence time • Confined or unconfined aquifer • Groundwater recharge rate	• Climate and season • Hydrological conditions (precipitation, runoff, flow, currents, velocity) • Geology and soil characteristics • Drainage characteristics • Temperature • Wind • Sunlight • Catchment area • Groundcover/ vegetation • Land use and management practices

Water pollutants are commonly characterised as point or diffuse, according to their source and pathway to the receiving environment:

- *Point sources* of pollution are directly discharged to receiving water bodies at a discrete location, such as pipes and ditches from sewage treatment plants, industrial sites and confined intensive livestock operations. The most severe water quality impacts from point source pollution typically occur during summer or dry periods, when river flows are low and the capacity for dilution is reduced, and during storm periods when combined sewer overflows operate more frequently. The "first flush" of a combined sewer system after a dry spell is particularly detrimental to surface water quality. Groundwater quality can also be affected where it interacts with polluted surface water.
- *Diffuse sources* of pollution (also referred to as non-point) are indirectly discharged to receiving water bodies, via overland flow (runoff) and subsurface flow (including pipeflow) to surface waters, and leaching through the soil structure to groundwater. Examples of diffuse pollution sources include nutrient runoff and leaching from the use of fertilisers in agriculture, atmospheric deposition of nitrogen oxides from energy and transport emissions, and runoff of petroleum hydrocarbons and heavy metals from urban surfaces not serviced by stormwater collection and treatment. The most severe water quality impacts from diffuse sources of pollution occur during storm periods (particularly after a dry spell) when rainfall induces hillslope hydrological processes and runoff of pollutants from the land surface.

The distinction between point and diffuse sources of pollution is also a function of policy and regulation. Point sources of pollution are largely under control in OECD countries because they are easier to identify and more cost-effective to quantify, manage and regulate. In comparison, diffuse sources are challenging to monitor and regulate due to: *i)* their high variability, spatially and temporally, making attribution of sources of pollution complex;

ii) the high transaction costs associated with dealing with large numbers of heterogeneous polluters (e.g. farmers, homeowners); and *iii)* because pollution control may require co-operation and agreement within catchments, and across sub-national jurisdictions and countries (OECD, 2012a). For these reasons, diffuse sources of pollution and their impacts on human and ecosystem health largely remain under-reported and under-regulated.

The damage caused by pollution disposal depends crucially upon the ecosystem's ability to absorb and dilute pollutants, which depends upon the ecosystem condition. If emissions exceed the assimilative capacity (absorptive or dilution capacity) of the system, they will accumulate and cause damage to the ecosystem. The deterioration of water quality has subsequent knock-on impacts on the functioning of in-stream invertebrates, fish, and aquatic plant communities (Doledec et al., 2006; Ling, 2010). This causes negative feedbacks, particularly the ability of ecosystems to process contaminants, thereby causing pollutants to accumulate in the environment and cause further damage. Conversely, activities that enhance ecosystems can increase their ability to process pollutants. Therefore, in addition to pollutants being classified as point source or diffuse source, pollutants can also be classified by the ability of the ecosystem to adsorb them (Lieb, 2004). The distinction below is relevant from a policy perspective:

- A *stock pollutant* is a pollutant with a long lifetime and for which the ecosystem has little or no absorptive capacity. Stock pollutants therefore accumulate in the environment. Examples include heavy metals, toxic contaminants, such as dioxins and polychlorinated biphenyls (PCB's), and non-biodegradable plastics. Groundwater aquifers, lakes, reservoirs and estuaries, particularly those with low recharge rates and high residence times, are examples of water bodies where their ability to absorb pollutants is limited. By their very nature, stock pollutants create interdependencies between decisions made today and the welfare of future generations, and the costs of treatment and damages typically rise over time (although advances in technology can reduce costs).
- Conversely, a *flow pollutant* has a short lifetime for which the ecosystem has some absorptive capacity. For example, suspended sediments washed out by rainfall into rivers only have a short lifetime. Organic pollution can be transformed into less-harmful inorganic matter by bacteria in water bodies, although this process uses up available oxygen and can cause loss of aquatic life. Nutrients (nitrates and phosphates) are required for aquatic plant growth, but in excess can proliferate aquatic weeds and turn waterways eutrophic. Since rivers are flowing, the concentrations of river pollutants decline more quickly than aquifers and lakes once pollution emissions have ceased. For this reason, river pollutants are generally short-lived and are often considered as flow pollutants. It is important to also note that a flow pollutant in one place, such as a river, can result in a stock pollutant elsewhere, such as an estuary, and as such, the source and dispersal of pollutants needs to be looked at systemically.

An overview of the main pollutants

The quality of water resources are affected by a number of pollutants. Table 1.1 summarises the most common pollutants and their sources, and they are individually described in more detail in Annex 1.A1. It is important to note that many of these pollutants may occur in parallel, and may be derived from a number of different sources and actors.

Table 1.1. **Water pollutants and their typical sources**

Pollutant	Media of origin[1]	Type of source[2]	Examples of source
Excess nutrient losses	L, W, A	P, D	Nitrogen and phosphorus fertilisers from agriculture and domestic lawns, livestock manure and slurry, and wastewater treatment plants. Nitrogen deposition from atmospheric sources of nitrogen oxides (NOx), ammonia (NH^3) and nitrous oxide (N_2O).
Microbial contamination	L, W	P, D	Pathogenic bacteria and viruses from wastewater treatment plants, combined sewer overflows, animal waste, septic tanks, land application of biosolids.
Acidification	L, W, A	D	Atmospheric pollutants (sulphur, nitrogen oxides, ammonia) and acid mine drainage.
Salinity	L, W	D	Irrigation of salt-affected soils, sea level rise and over-abstraction of groundwater in coastal areas, de-icing salts used on roads.
Sedimentation	L, W	P, D	Erosion of topsoil and peatlands, livestock manure spreading on pasture, sediment release from dams, wastewater treatment plants, food processing waste.
Toxic contaminants	L, W	P, D	Pesticides and herbicides for plant and animal protection in agriculture, roadside and domestic use of herbicides. Heavy metals[3] from urban stormwater runoff, land application of biosolids, mining waste, industrial waste, and aging and corroding infrastructure. Natural arsenic groundwater pollution. Chlorinated solvents and other chemicals from transport, industry, spills, fracking, urban stormwater runoff and leaking storage tanks.
Thermal pollution	L, W	P, D	Warm water from urban stormwater runoff, and power plants and industrial manufacturers that use water as a coolant. Cool water from dam releases.
Plastic particle pollution	L, W	D	Rubbish dumping by individuals, the plastic production industry, recreational and commercial fishers and urban stormwater runoff.
Contaminants of emerging concern (CECs)	W	P	Commonly sourced from the household (through wastewater treatment plants), and to a lesser extent, from agriculture. Examples include pharmaceuticals, antibiotics, hormones, personal care products, perfluorinated compounds, flame retardants, plasticizers, detergent compounds, caffeine, fragrances, cyanotoxins, engineered nanomaterials, anti-microbial cleaning agents and their transformation products.

Notes: 1. Land (L), Air (A), Water (W); 2. Point source (P), Diffuse source (D); 3. The most common heavy metals are cadmium, mercury, lead, arsenic, manganese, chromium, cobalt, nickel, copper, zinc, selenium, silver, antimony and thallium.

Negative feedbacks on water quality

Other factors that contribute to degradation of freshwater ecosystems, and thus their ability to process contaminants, include the introduction of invasive alien species and anthropogenic geomorphological modifications to river systems. According to the IUCN, invasive alien species constitute the second most severe threat to freshwater fish species (Darwall et al., 2009), and the spread of invasive alien species is projected to increase due to a combination of increasing trade and climate change (Death et al., 2015; Rabitsch et al., 2016; Walther et al., 2009).

Changes in the natural geomorphology and flow of water bodies (e.g. channelised rivers, dams, canals, flood defences) can also have some effects on water quality and the ability of ecosystems to process and retain pollutants (Nilsson and Malm Renöfält, 2008; Wagenschein and Rode, 2008). For example, a study on the Weisse Elster River, Germany, revealed that the nitrogen retention rate is almost 2.4 times higher in a natural section of the river compared with a heavily modified and channelised section (Wagenschein and Rode, 2008).

Links between water quality and water quantity

Water quality and quantity are inextricably linked. Water pollution reduces the quantity of useable water and therefore exacerbates the problem of water scarcity. Water scarcity and droughts reduces the capacity for dilution of point source discharges to surface waters, and additional treatment of wastewater may be required to compensate for the lower dilution capacity of water bodies. Water scarcity also increases water temperatures which can affect freshwater ecosystems and nuisance algal growth. Conversely, high rainfall events and flooding induce diffuse pollution from land runoff (agricultural and urban) and trigger combined sewer overflows into rivers.

There can be competing demands for quality and quantity, driven by the requirements of the users. Different users require different volumes of water at different times and places, and different users are more or less sensitive to water quality. They also require varying levels of certainty regarding the availability and quality of water, and citizens have increasing expectations as regards the quality of water. There may be trade-offs and co-benefits between water quantity and quality management, and other important sectoral policies, such as land, energy, biodiversity, urban planning, health care, waste, construction, transport, and climate change (discussed in Chapter 4).

Ongoing challenges of diffuse pollution sources and eutrophication in OECD countries

Eutrophication and harmful algal blooms in freshwater systems are quickly becoming a global epidemic. For instance, there have been reports of algal blooms in Lake Nieuwe Meer in The Netherlands (e.g. Johnk et al., 2008), Lake Erie in North America (e.g. Michalak et al., 2013), Lake Taihu in China (e.g. Qin et al., 2010), and Lake Victoria in Africa (e.g. Sitoki et al., 2012). Furthermore, the effects of climate change are expected to exacerbate existing eutrophication and algal bloom problems (Bates et al., 2008).

Figure 1.3 illustrates that the main source of nutrient loading in OECD countries is from agriculture, and this is largely because the methods and policies driving the "green revolution"[5] frequently lacked incentives for prudent use of inputs and promoted expansion of cultivation into areas that could not sustain high levels of intensification (Pingali, 2012) (advances in wastewater treatment have also increased the proportion of nutrient pollution from agriculture).

In Europe, nutrient pollution, leading to eutrophication, is a widespread problem which occurs in about 30% of water bodies in 17 member States (European Commission, 2012b). The latest assessments on the implementation of the Water Framework Directive 2000/60/EC (WFD), as well as studies carried out in the framework of international conventions, show that diffuse sources of pollution are the greatest obstacle to achieving "good" status in EU waters (European Commission, 2013). Agriculture remains the predominant source of reactive nitrogen discharged into the environment, and a significant source of phosphorus, mainly from livestock manure and fertilisers (European Commission, 2013).

In the midst of generally improving farm practices, there remain "hotspots" where water quality improvements are not yet forthcoming. For example, in the European Union between 2008 and 2011, almost 15% of groundwater monitoring stations exceeded 50 mg nitrate per litre (the WHO standard for nitrates in drinking water), and approximately 30% of river monitoring stations and 40% of lake monitoring stations were eutrophic or hypertrophic (European Commission, 2013). Some EU countries[6] have been convicted of failing to fulfil their obligations to the European Commission Nitrates Directive (91/676/EEC). In each case, the European Court of Justice has ordered member states to strengthen their regulations to comply with the Nitrates Directive.

Figure 1.3. **Percentage share of agriculture in total emissions of nitrates and phosphorus in surface water, OECD countries, 2009 or latest available year**

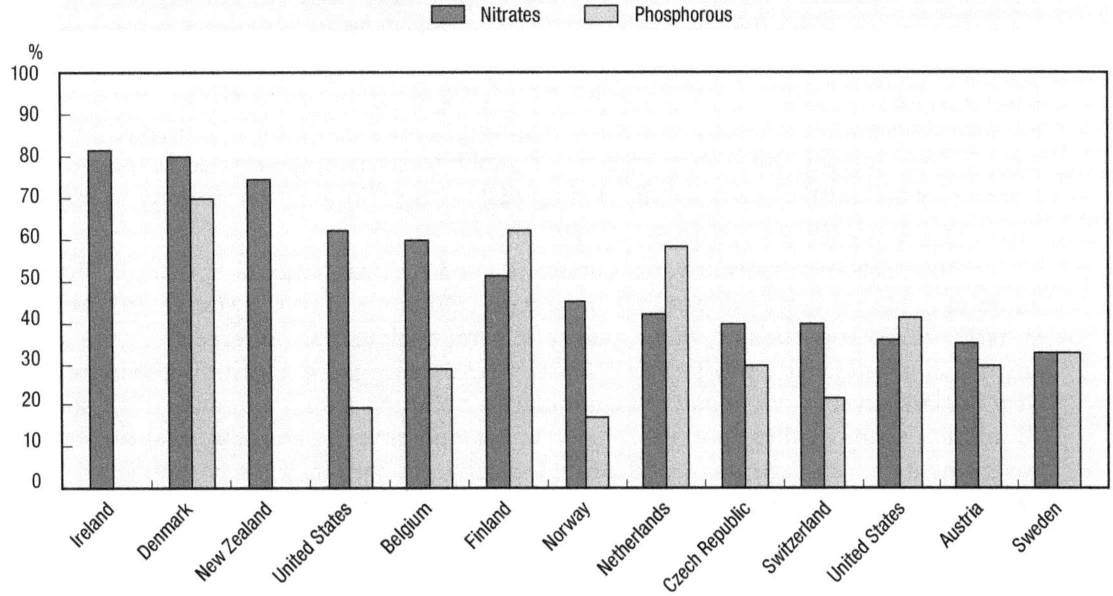

Notes: Countries are ranked in descending order of highest share of nitrates in surface water. For nitrates, the figures presented correspond to the year 2000 for Austria, Czech Republic, New Zealand, Norway, Switzerland and United States; 2002 for Denmark; 2004 for Finland and Ireland; 2005 for Belgium (Wallonia); 2008 for United Kingdom; and 2009 for Netherlands and Sweden. For phosphorous, the figures presented correspond to the year 2000 for Austria, Czech Republic, Norway, Switzerland and United States; 2002 for Denmark; 2004 for Finland; 2005 for Belgium (Wallonia); and 2009 for Netherlands, Sweden and United Kingdom.

Source: OECD (2013), OECD Compendium of Agri-environmental Indicators, OECD Publishing. http://dx.doi.org/10.1787/9789264181151-en.

StatLink http://dx.doi.org/10.1787/888932793015

Similar eutrophication problems have been reported in North America's Great Lakes, largely due to high phosphorus loading. For example, the water quality of Lake Erie (bordering the States of New York, Pennsylvania, Ohio and Michigan, and the Canadian province of Ontario) has been an ongoing concern in relation to nutrient overloading from fertilisers, and human and animal waste, leading to eutrophication, hypoxia and algal blooms. Despite some initial improvement in response to the 1972 Great Lakes Water Quality Agreement, and a reduction in phosphorus from sewage treatment plants and other point sources, diffuse sources from agriculture and domestic lawns have remained largely unaccounted for, and since the mid-1990s, Lake Erie has been returning to a more eutrophic state (Scavia et al., 2014).

In 2014, the eutrophication of Lake Erie resulted in a seven-day tap water ban for Toledo, Ohio when blooms of toxic algae shut down drinking water supplies from the lake, affecting more than 400 000 people, and closing local restaurants, universities and public libraries (Circle of Blue, 2014). Furthermore, the water ban occurred after the city of Toledo increased spending on water treatment chemicals - USD 4 million in 2013; double what it spent in 2010. Further upgrades estimated at USD 321 million are needed for the city's treatment plant; costs that are to be met by the tax payer. It is estimated, that in order to reduce eutrophication and the central basin hypoxic area to levels observed in the early 1990s, total phosphorus loading will need to be reduced by 46% from the 2003–2011 average (Scavia et al., 2014). In acknowledgement of the ongoing water quality problems, the hypoxia-based loading targets were revised in the 2012 Great Lakes Water Quality Agreement, and in 2016 the governments of Canada and the United States announced bi-national phosphorus load reduction targets of 40% for Lake Erie.

In the United Kingdom (UK), water quality has improved as a result of a major investment programme focusing on point source pollution from industrial discharges and wastewater treatment plants. However, diffuse pollution, coupled with remaining point source pollution, mean that approximately 15% of the urban river network in England and Wales fall into the poor or bad categories of the WFD (Royal Commission on Environmental Pollution, 2007). Furthermore, at least 50% of UK groundwater used for public supply is showing significant deterioration in quality (Royal Geographical Society, 2012; UK Environment Agency, 2015) with sources of public water supply affected (although well-treated) by agricultural pollution (including historical pollution), particularly in terms of high levels of nitrates and pesticides (Water UK, 2013).

In New Zealand, overall water quality is good by international standards, but this varies around the country depending on land use, climate and geology (MfE, 2013; MfE and StatsNZ, 2015). In particular, water quality in some regions has suffered from the steady expansion of intensive dairy farming (OECD, 2015a). Nitrogen is continuing to increase in New Zealand rivers – the result of accumulative pollution from rural and urban sources. To address water quality concerns, the New Zealand government now requires regional governments to manage point and diffuse discharges within set environmental limits (MfE, 2014a). Results have already started to show improvement, with some water bodies making significant recoveries, such as the Rotorua Lakes (MfE, 2014b). There have also been improvements in phosphorus levels in rivers due to riparian planting, reduced phosphorus fertiliser-use and soil conservation efforts over the past 10-20 years (MfE and StatsNZ, 2015).

In Chile, significant progress has been made in providing improved sanitation in both urban and rural areas such that 99% of the population now have access to improved sanitation (WHO/UNICEF, 2015) and nearly 70% are connected to a public wastewater treatment plant with secondary or tertiary treatment (OECD, 2013a). However, there are two ongoing concerns regarding water quality in Chile. Firstly, diffuse pollution from agriculture is of concern with high levels of nitrates and pesticides observed in surface water. Secondly, mining and other industrial activities, mainly in northern and central Chile, are major sources of pollution. It is estimated that over 60 % of industrial discharges (including tailings) flows into sewerage networks, mixes with domestic sewage and is deposited in the river basins and irrigation channels, or is discharged to the soil or directly into the sea. This is of particular concern, especially in regions where water flows for dilution of acidity, hazardous chemicals and heavy metals are small or non-existent.

In Japan, there has been a significant reduction in heavy metals in recent years owing to tighter regulations on industrial wastewater. Environmental quality standards for organic pollution and nutrients are not being met in approximately 10% of Japan's water bodies. In particular, there has been little improvement in enclosed water areas such as inland seas, inlets, lakes, and reservoirs (Government of Japan, 2015a). Eutrophication occurs in a large number of Japanese lakes and reservoirs, many of which are used for municipal and industrial water supply (Government of Japan, 2015b). As a result, algal blooms are frequent and disrupt water treatment facilities.

Korea has invested in water infrastructure over the last 50 years, reaching a high level of access to water supply and sanitation services with 90% of the population connected to a public wastewater treatment plant with secondary or tertiary treatment (OECD, 2013a). Regulations and economic instruments have been implemented to manage point source pollution and improve water quality since the early 1990s. However, there are ongoing challenges with nutrient pollution in the four major rivers with the occurrence of eutrophication and frequency of algal blooms increasing. It is estimated that diffuse pollution sources (from both urban and rural areas) were responsible for 68% of the total pollutant loading in rivers in 2010. This is projected to reach 72% by 2020, primarily due to urbanisation and an increase in paved impervious areas and stormwater runoff (Ahn, 2015). Korea's Second Comprehensive Nonpoint Pollution Source Control Measure (2012-2020) aims to reduce diffuse sources of BOD and total phosphorus in the four major rivers by 24.6% and 22.5% respectively, by 2020.

In summary, OECD countries face significant challenges regarding the control of diffuse pollution sources, with the most prevalent water quality problem being nutrient loading and eutrophication. In addition to the challenging nature and associated costs of diffuse pollution management, there are also ecosystem delays (the time difference between implementation of abatement measures and actual measurable effects) due to the long-time scales of eutrophication (Gustafsson et al., 2012). Ecosystem responses to measures that reduce eutrophication illustrate that feedbacks and climate change impacts can keep ecosystems in a certain state and cause delays of decadal scale in ecosystem response (Varjopuro et al., 2014). These factors are illustrated in the case of the Baltic Sea (Box 1.1).

Box 1.1. **Ecosystem delays and ongoing nitrogen and phosphorus pollution of the Baltic Sea**

The Baltic Sea has acted as sink for much of the nutrient loss from agricultural diffuse pollution sources from Scandinavia, Finland, the Baltic countries, and the North European Plain. Levels of nitrogen and phosphorus are four and eight times greater than what they were in the early 1900s (WWF, 2015). In order to achieve good ecological status under the EU Water Framework Directive, it is estimated that phosphorous and nitrogen inputs to the Baltic Sea need to be reduced by about 42% and 18%, respectively (Skogen et al., 2014). In response, the HELCOM Baltic Sea Action Plan was launched; an ambitious programme to restore the good ecological status of the Baltic marine environment.

However, despite a reduction in nutrient loading in recent years, little change in the eutrophic effects of the Baltic Sea has been observed (WWF, 2015). Furthermore, simulations indicate that no future improvement in the water quality of the Baltic Sea can be expected from the decrease in nutrient loads in recent decades (Skogen et al., 2014, Gustafsson et al., 2012). This is for three reasons:

1. The time scales of eutrophication are exceptionally long and all efforts taken to reduce nutrient loads up to now have basically resulted in maintaining the status quo (Gustafsson et al., 2012);
2. Climate change is stimulating eutrophication as higher temperatures in the Baltic Sea region increases the growth and decomposition rates of the algae, thereby enhancing oxygen depletion and counteracting land practices that reduce eutrophication (WWF 2015, Lennartz et al., 2014); and
3. The expected development of agriculture in the new EU countries around the Baltic Sea will worsen the conditions measurably if no additional action is taken to reduce the harmful effects of nutrient losses (WWF, 2015, Gustafsson et al., 2012).

Sources: Gustafsson et al. (2012); Lennartz et al. (2014); Skogen et al. (2014); WWF (2015).

Water quality and climate change

Anthropogenic warming of the climate system is "unequivocal" (IPCC, 2014a). Concentrations of greenhouse gases have increased to unprecedented levels resulting in warming of the atmosphere and ocean, reductions in snow and ice, sea level rise, ocean acidification, changes in the global water cycle, and changes in climate extremes (IPCC, 2014a).

The IPCC (2014a) predicts that climate change will have significant additional impacts on existing water quality challenges, due to altered precipitation and flow regimes, altered thermal regimes, and sea level rise. With a "high level of confidence", many forms of water pollution will be exacerbated – from sediments, nutrients, dissolved organic carbon, pathogens, pesticides and salt, as well as thermal pollution, with possible negative impacts on freshwater ecosystems, human health, and water system reliability and operating costs (Bates et al., 2008). The interaction of increased temperature; increased sediment, nutrient and pollutant loadings during heavy rainfall, runoff and soil erosion; increased concentrations of pollutants during droughts; and disruption of treatment facilities during floods, will reduce raw water quality and pose risks to drinking water quality even with conventional treatment (IPCC 2014a; Delpla et al., 2009). For example, under a drier future (projected by the CSIRO global circulation model), coupled with medium levels of income and population growth, the number of people living in environments with high water quality risks[7] due to excessive nitrogen and phosphorus loading will raise to one-third of the global population by 2050 (172% and 129% increase for nitrogen and phosphorus respectively), and for BOD, one-fifth of the population (144% increase) (IFPRI and Veolia, 2015).

Sea-level rise is projected to extend areas of estuaries and increase salt-water intrusion of freshwater aquifers, resulting in a decrease of freshwater availability for humans and ecosystems in coastal areas (Bates et al., 2008). All of these climate change induced alterations in temperature and flow regimes, water quality and salinity, will lead to shifts in freshwater species distributions, reduced ecosystem functioning and further exacerbate existing water quality problems (IPCC, 2014b). A summary of the effects of climate change on water quality is presented in Table 1.2.

Water quality issues were identified as a main concern in 15 OECD countries in the report on *Water and Climate Change Adaptation* (OECD, 2013b). For example, in Canada, warmer conditions will increase surface water temperatures, decrease the duration of ice cover and lower water levels, and is projected to result in higher pollutant concentrations. In addition, increased flooding is also expected to contribute to water quality degradation. Korea anticipates an increase in the risk of algal outbreaks in public waters due an increase in water temperature and changes in rainfall patterns. They also expect an increase in the risk of water quality degradation due to diffuse source pollution resulting from an increase in frequency and intensity of high rainfall events. In Chile, surface water quality is expected to decline due to increased flooding and storm events, and reduced capacity for dilution during droughts. Further groundwater salinisation and pollution is anticipated in coastal zones and northern areas of Chile. Denmark, the EU, Japan, Mexico, and the Netherlands are also concerned about groundwater salinisation associated with sea level rise, reduced groundwater recharge and increased demand for irrigation during the dry season (OECD, 2013b).

The effects of climate change at the local level should be interpreted cautiously, considering the type of water body, the pollutant of concern, the hydrological regime, and the many other factors identified in Figure 1.2 (Whitehead et al., 2009). In general, current information about the water quality impacts of climate change is limited, including their socio-economic dimensions (Bates et al., 2008). There is a need to improve understanding and modelling of the impacts of climate change on water quality at scales relevant to decision making, and of vulnerability to and ways of adapting to those impacts (IPCC, 2014a). Management approaches need to account for uncertainties around climate change projections regionally and locally, and the impacts on water quality.

Table 1.2. **Effects of climate change on water quality**

Direct climate change impacts and indirect effects on water quality			
Increased severity and frequency of flooding	**Increased severity and frequency of droughts**	**Sea level rise**	**Increased water temperature**
Disruption of treatment facilities during floods, subsequent risks to drinking water quality and human health (e.g. infectious diseases). Increased runoff and nutrient loading leading to increased eutrophication. Increased runoff and greater loads of heavy metals, salts and other pollutants. Increased soil erosion, sediment, organic matter and pathogens loadings, subsequent impairment of conventional drinking water treatment. Increased release of combined sewer overflow. Increased re-suspension of riverbed and lakebed sediments containing high metal concentrations, associated contamination of water and drinking water risks, transfer of contaminated sediments to floodplain soils used for agriculture. Impacts on freshwater ecosystems: extinction and shifts in distribution of species, changes in river geomorphology and habitat, increased dispersal of invasive species, reduced functioning of ecosystem services.	Reduced dilution of pollutants from point sources as a result of a reduction in rainfall, groundwater recharge and glacier retreat. Soil shrinking and damage/cracking of water infrastructure, subsequent risks to drinking water quality and environment, and increased maintenance costs. Increased severity and frequency of forest wildfires, increase in erosion and reduced filtration/regulation ecosystem services affecting water quality. Impacts on freshwater ecosystems: extinction and shifts in distribution of species, reduced functioning of ecosystem services.	Extension of estuaries and salt water intrusion of groundwater aquifers, especially in areas where rainfall (and recharge) is expected to decline and water demand to increase. Increased treatment costs for drinking water use, industrial production and agriculture. Intrusion of saline water to sewers, subsequent increase of corrosion and maintenance of water infrastructure. Impacts on freshwater ecosystems: extinction and shifts in distribution of species, loss/reduced functioning of ecosystem services.	Reduced solubility of oxygen, higher metabolism, and increased stratification of the water column, resulting in increased hypoxia, algal blooms and associated toxins, with subsequent risks to drinking water quality and recreational use. Increase of the growth and survival of pathogens, risks to drinking water quality and human health (infectious diseases). Impacts on freshwater ecosystems: extinction and shifts in distribution of species, loss/reduced functioning of ecosystem services. Increase in soil erosion associated with melting of permafrost.

Sources: Bates et al. (2008); Death et al. (2015).

Water quality challenges for cities of OECD countries

The impacts on water quality, whether rural or urban in source, largely fall on cities, where the value of assets at risk is concentrated. Future population growth, urbanisation and more stringent standards (such as those imposed under the WFD), will place extra demands on existing systems and mean that significant investment in drinking water and wastewater treatment infrastructure are required in order to prevent water-related disease outbreaks and not place additional nutrient, pathogenic and organic loads in river systems. Furthermore, as our understanding of contaminants of emerging concern (CECs) and their effects on human and environmental health improves, future regulations may require treatment to remove CECs (conventional water purification and wastewater treatment plants are not effective at removing CECs).

Continual control of point source pollution is essential for water quality. However, some countries have reached the economic limit in terms of public water supply and sewerage connection and must find other ways of serving small, isolated settlements (OECD, 2011). Decentralised water and wastewater systems, and water fit for purpose, are potential solutions to this problem (see OECD, 2015b). Without effective wastewater treatment plants and sewerage systems, wastewater effluent can add to the existing problems of diffuse pollution from agriculture in the following ways:

- *Wastewater effluent without the removal of nitrogen and phosphorus* (i.e. from primary or secondary wastewater treatment plants[8]) adds to the concentration of nutrients in receiving water bodies from diffuse sources of pollution. The problem is exacerbated during dry periods when a reduction in river flow decreases the capacity of river systems to dilute wastewater effluent discharges;

- *Combined sewer overflows* discharge untreated wastewater directly into watercourses during storm events, when the storage and treatment capacity of wastewater treatment plants are exceeded. A classic example is the River Thames in London which essentially acts as an open sewer during periods of significant rainfall (Box 1.2).
- More modern separate sewer systems isolate wastewater from stormwater for separate treatment. However, during extended storm events, when the stormwater storage and treatment capacity of wastewater treatment plants are exceeded, *untreated (but screened) stormwater* and associated contaminants are discharged directly to the receiving environment. In addition, there are reported occurrences of *accidental cross connection of pipes* which result in the direct discharge of untreated wastewater to the environment. For example, cross-connections have been recognised as a problem in the United Kingdom (Royal Commission on Environmental Pollution, 2007);
- *Leakages from aging sewer infrastructure, and lack of maintenance,* contribute to diffuse pollution of groundwater; and
- *Dumped or landfilled sewage sludge* can leach nutrients, pathogenic organisms and heavy metals to groundwater (Van Den Berg, 1993). *Land application of sewage sludge and irrigation with wastewater* that is not adequately treated can also contribute to surface runoff and leaching of nutrients, pathogens and heavy metals.

Box 1.2. **Combined sewer overflows contribute to diffuse pollution: Example of the London sewer system**

The London sewer system discharges untreated sewage and diluted stormwater to the River Thames, on average, once per week (Thames Tideway Tunnel, 2015). This is because the existing infrastructure, now 150 years old and designed for a maximum capacity of 4 million people, can no longer cope with the stresses of serving 8 million people and the change in weather patterns associated with climate change. In order to protect the River Thames from increasing pollution, and to meet European environmental standards, a major new sewer - The Thames Tideway Tunnel – will be constructed at considerable cost (GBP 4.1 billion) to intercept current overflow discharge points in the system and transfer the sewage to Beckton Sewage Treatment Works for treatment before discharge (a case study on financing the Thames Tideway Tunnel is presented in Chapter 3).

Source : Thames Tideway Tunnel (2015).

The effects of climate change on water quality detailed in the prior section will mean water and sanitation utilities will be confronted by further economic and operational challenges requiring additional or new treatment facilities and technologies:

- *Higher water temperatures* will stimulate more algal blooms and increase human health risks from cyanotoxins and natural organic matter in water sources (IPCC, 2014b). Temperature increases and precipitation pattern changes associated with climate change are also predicted to increase the growth, survival, and transport of enteric bacteria (Liu et al., 2013) and therefore increase the risk of water-borne diseases ["very high confidence"] (IPCC, 2014b). This will require additional or new treatment of drinking water. On the plus side, warmer water can increase biological reactions in drinking and wastewater treatment, particularly biological nitrogen removal, thereby potentially reducing treatment costs (Kadlec and Reddy, 2001). Conversely, cooler water from increased snow and glacier melt can have the opposite effect (Plósz et al., 2009).

- *Drier conditions* will increase pollutant concentrations, due to reduced environmental flows and dilution capacity (IPCC, 2014b), and therefore may require effluent to be treated to a higher quality. The risk of contamination of water supplies will also increase in response to reduced dilution of upstream pollution and potential increases in water-related disease outbreaks, harmful algal blooms and other health effects. Wastewater reuse will increasingly be a cost-effective alternative of conventional water supply. Soil shrinking due to reduced soil water content may induce cracking of water mains and sewer pipes, making them vulnerable to infiltration and exfiltration of water and wastewater. The combined effects of warmer temperatures, increased pollutant concentrations, longer retention times, and sedimentation of solids may lead to increasing corrosion of sewers, shorter asset lifetimes, increased risk of drinking water pollution, and higher maintenance costs (IPCC, 2014b).
- *Wetter conditions* will increase runoff, which increases loads of pathogens, nutrients, and suspended sediment (IPCC, 2014b), particularly following a dry period, and increases the risk of combined sewer flooding, water-related disease outbreaks, harmful algal blooms and other health effects. The maximum loading and capacity of wastewater treatment plants may need to be increased, and overflow infrastructure adapted, to cope with increased volumes of wastewater in short periods (Plósz et al., 2009). Increased storms, floods and sea level rise may be harmful to infrastructure, particularly given that wastewater treatment plants are often located in low-lying, coastal areas. Rising downstream water levels may make pumping drinking water and effluent a requirement, increasing energy needs and costs.
- *Sea level rise* will increase the salinity of coastal aquifers, in particular where groundwater recharge is also expected to decrease (IPCC, 2014b). This will require additional or new treatment facilities to treat water for potable consumption, and increased maintenance to reduce the effects of infrastructure corrosion associated with high salinity. High salinity may also have consequences for industrial production and agriculture as water quality standards are exceeded (Zwolsman et al., 2011). Sea level rise and strong waves during storms may endanger the location of wastewater treatment plants in low-lying coastal areas.
- *Reliance on green infrastructure and ecosystem services*, such as regulating ecosystem services provided by forested catchments and wetlands, may be jeopardised with increased forest wildfires, pest and disease outbreaks, increased tree mortality and other indirect effects of climate change (such as land use change and increased irrigation for food security) (Smith et al., 2011). Further investments in the protection and conservation of green infrastructure and natural capital may complement conventional grey infrastructure and may be more cost-effective than conventional grey infrastructure alone.

Challenges also remain regarding the upgrade of ageing water supply and sewage systems (OECD, 2014; OECD, 2015b). For example, in some parts of the United Kingdom, sewerage systems are approaching 200 years old (Royal Commission on Environmental Pollution, 2007). In the city of Flint, Michigan, United States, a contaminated public water supply, ageing infrastructure and inadequate maintenance of the city's water distribution network were part of what caused the Flint Water Crisis (Box 1.3). The case study highlights the importance of historic pollution, financing and investment in water infrastructure and maintenance, compliance with water quality standards, and transparency and communication to the public. Drinking water risk assessments can help identify and prioritise where interventions (e.g., water source protection, wastewater treatment upgrades, water distribution system repairs or replacements, and/or optimisation of filtration and disinfection) are required to reduce risks (DeFelice et al., 2015).

Box 1.3. The Flint Water Crisis, Michigan, United States

The Flint water crisis of 2014-2015 was the result of a series of governance and infrastructure failures that resulted in drinking water being contaminated with lead and associated ill health effects to the city's 100 000 residents. The crisis provides a number of lessons regarding the political cost of deferring critical infrastructure investments and prioritising economic concerns over the provision of clean, safe water.

The Flint authorities switched the public water supply in April 2014 from Lake Huron (treated and supplied by the Detroit Water and Sewerage Department), to the local Flint River, which had not been used for consumption since the early 1960s because of high industrial pollution. The decision to switch was made in an attempt to obtain more affordable water rates for residents, 40% of which live below the poverty line. However a series of problems was associated with this switch to the Flint River source:

- The water from Flint River required significant chemical treatment before distribution which subsequently caused corrosion of ageing lead pipes which led to extremely elevated levels of lead in drinking water. Officials failed to apply corrosion inhibitors.
- Residents began complaining about the colour, taste and odour of public water supply almost immediately.
- High levels of chlorine used to disinfect the drinking water, in combination with the organic matter present in the supply, resulted in elevated levels of Trihalomethanes (with which long term exposure has been linked to cancer and other diseases) in August 2014. A violation notice by the Michigan Department of Environmental Quality (DEQ) was issued to the city in January 2015.
- The first indication of any corrosion was with a General Motors plant in Flint complaining that the water was corroding car parts. It stopped using Flint water in October 2014.
- In February 2015 the first independent studies were released showing lead contaminated drinking water and elevated levels of lead in children.
- EPA officials warned the state DEQ repeatedly, beginning in February 2015, that the lack of corrosion control in Flint water mains would lead to a serious lead safety hazard in drinking water supplies.
- In October 2015, the Governor admitted the situation was far graver than he initially understood and announced a USD 12 million plan to transfer Flint back to its previous supply with the city of Detroit.

The above problems resulted in a state of emergency declared by the Governor on 5 January 2016, and a federal state of emergency declared by President Obama on 16 January 2016. Researchers estimate between 6 000 and 12 000 children have been exposed to extremely high levels of lead that has the potential to cause irreversible health and neurological problems. As such, the Flint water crisis will have long term impacts associated with the public trust of government official and regulators, and long term health costs to the residents of Flint.

The evidence is mounting that federal, state and local officials ignored or neglected indicators of a growing water crisis. A number of investigations have been opened, several government officials have resigned over the mishandling of the crisis, and a number of lawsuits have been filed against government officials.

Sources: AWWA (2012); Circle of Blue (2016); Fisher (2016); The Guardian (2016); The Guardian (2015); USA Today (2016); Walton (2016).

Notes

1. Pollution is due to the influence or activities of people. Water contamination may be natural or caused by pollution (anthropogenic).
2. Pollution "hotspots" are specific locations that are identified as suffering from high pollution, or most likely to be subject to water pollution risks in the future, due to higher hazard, exposure and/or vulnerability.
3. Climate change is projected to increase water temperatures and precipitation intensity, and induce longer periods of low flows which will "exacerbate many forms of water pollution, including sediments, nutrients, dissolved organic carbon, pathogens, pesticides, salt and thermal pollution" (Bates et al., 2008).
4. Biochemical Oxygen Demand (BOD) is an indicator of the total load of organic matter in a water body. A high BOD reduces available supply of dissolved oxygen and causes mortality of aquatic organisms.
5. A significant increase in agricultural productivity beginning in the 1940s and resulting from the introduction of high-yield varieties of grains, the use of irrigation, fertilisers and pesticides, and improved farm management techniques.
6. In recent years, France, Greece, Poland and Luxembourg have been taken to court over nitrate pollution (European Commission 2015), and Estonia has been warned (European Commission, 2016).
7. High pollution risk is defined as adverse impacts on humans, the environment, and the economy are likely to occur. These figures are conservative as populations living in basins without water quality data are excluded.
8. Effective secondary treatment typically removes 85% of the suspended solids and BOD, and some heavy metals. When coupled with a disinfection step, these processes can provide substantial, but not complete, removal of bacteria and viruses. Secondary treatment removes little phosphorus, nitrogen, non-biodegradable organics, or dissolved minerals. Tertiary (advanced) treatment is required to remove more than 99 % of all the impurities from sewage (including nitrogen and phosphorus), producing an effluent of almost drinking-water quality. Advanced treatment processes are sometimes combined with primary or secondary treatment to remove phosphorus (FAO, 1992; World Bank, 2015).

References

Ahn, J-H. (2015), Improvement of services in the Republic of Korea, case study: Water, Sanitation and Hygiene, (WASH), 2015 UN-Water Annual International Zaragoza Conference. Water and Sustainable Development: From Vision to Action, 15-17 January 2015, www.un.org/waterforlifedecade/waterandsustainabledevelopment2015/pdf/Jong_Ho_Ahn_KoreaGDG.pdf.

Allan, C., Xia, J. and C. Pahl-Wostl, (2013) Climate change and water security: challenges for adaptive water management. *Current Opinion in Environmental Sustainability*, Vol. 5, No. 6, pp. 625–632.

AWWA (2012), Buried no longer: Confronting America's Water Infrastructure Challenge, American Water Works Association, Denver, CO.

Bates, B.C., Z.W. Kundzewicz, S. Wu and J.P. Palutikof, Eds. (2008), Climate Change and Water. Technical Paper of the Intergovernmental Panel on Climate Change, IPCC Secretariat, Geneva, 210 pp.

Circle of Blue (2016), Flint Water Crisis: Infrastructure, Economy, Health, and Politics, www.circleofblue.org/flint/ *[accessed 29/03/2016]*.

Circle of Blue (2014), Toledo Issues Emergency 'Do Not Drink Water' Warning to Residents, Saturday, 02 August 2014.

Darwall, W. et al. (2009), Freshwater biodiversity: a hidden resource under threat, *In*: Vié, J.-C., Hilton-Taylor, C. and S.N. Stuart (eds.), Wildlife in a Changing World – An Analysis of the 2008 IUCN Red List of Threatened Species, IUCN, Gland, Switzerland 180 pp.

Death, R.G., Fuller, I.C. and M.G. Macklin (2015), Resetting the river template: the potential for climate-related extreme floods to transform river geomorphology and ecology, *Freshwater Biology*, Vol. 60, pp. 2477–2496.

DeFelice, N.B., Johnston, J.E. and J. MacDonald Gibson (2015), Acute Gastrointestinal Illness Risks in North Carolina Community Water Systems: A Methodological Comparison, *Environmental Science and Technology*, E-publication ahead of print 13 July 2015.

Delpla, I., Jung et al. (2009), Impacts of climate change on surface water quality in relation to drinking water production, *Environment International*, Vol. 35, pp. 1225–1233.

Doledec, S., Phillips, N., Scarsbrook, M., Riley, R. H. and Townsend, C. R. (2006) 'Comparison of structural and functional approaches to determining landuse effects on grassland stream invertebrate communities', Journal of the North American Benthological Society, 25(1), 44-60.

European Commission (2013), Report from the Commission to the Council and the European Parliament on the implementation of Council Directive 91/676/EEC concerning the protection of waters against pollution caused by nitrates from agricultural sources based on Member State reports for the period 2008–2011, European Commission, Brussels, Belgium.

European Commission (2012a), Attitudes of Europeans towards water – related issues: Summary, Flash Eurobarometer Series No. 344, European Commission, Brussels, Belgium.

European Commission (2012b), Blueprint to Safeguard Europe's Water Resources, European Commission, Brussels, Belgium.

FAO (1992), Wastewater treatment and use in agriculture, FAO Irrigation and Drainage Paper.

Fisher, R. (2016), Mich. infrastructure negligence evident in Flint crisis, Detroit Press Free, www.freep.com/story/opinion/contributors/2016/02/06/michigan-infrastructure-flint-crisis/79642266/ *[accessed 29/03/2016]*.

Gallup Poll (2014), "Americans Show Low Levels of Concern on Global Warming", www.gallup.com/poll/168236/americans-show-low-levels-concern-global-warming.aspx.

Government of Japan (2015a), Public waters water quality measurement results (平成27年度公共用水域水質測定結果), [online], http://www.env.go.jp/water/suiiki/ [accessed 20/02/2017].

Government of Japan (2015b), State of Japan's Environment at a Glance: Japanese Lake Environment - The Central Issue-Eutrophication, [online], www.env.go.jp/en/water/wq/lakes/issue.html [accessed 10/07/2015].

Gustafsson, B. G. et al. (2012), Reconstructing the development of Baltic Sea eutrophication 1850-2006. *AMBIO*, Vol. 41, pp. 534-548.

Hanley, N., Shogren, J.F. and E. White (2013), The Economics of Water Pollution, In: *Introduction to Environmental Economics*, second Edition, Oxford University Press, Oxford, UK.

Hill, S. (2013), "Reforms for a Cleaner, Healthier Environment in China", *OECD Economics Department Working Papers*, No. 1045, OECD Publishing, Paris, http://dx.doi.org/10.1787/5k480c2dh6kf-en.

Hughey, K.F.D., Kerr, G.N. and Cullen, R. 2013. Public Perceptions of New Zealand's Environment: 2013. EOS Ecology, Christchurch.

IFPRI and Veolia (2015), The murky future of global water quality: New global study projects rapid deterioration in water quality, International Food Policy Research Institute (IFPRI), Washington, D.C. and Veolia Water North America, Chicago, IL.

IPCC (2014a), Climate Change 2014: Synthesis Report. Contribution of Working Groups I, II and III to the Fifth Assessment Report of the Intergovernmental Panel on Climate Change [Core Writing Team, R.K. Pachauri and L.A. Meyer (eds.)]. IPCC, Geneva, Switzerland, 151 pp.

IPCC (2014b), Climate Change 2014: Impacts, Adaptation, and Vulnerability. Part A: Global and Sectoral Aspects. Contribution of Working Group II to the Fifth Assessment Report of the Intergovernmental Panel on Climate Change [Field, C.B. et al. (eds.)]. Cambridge University Press, Cambridge, United Kingdom and New York, NY, USA, 1132 pp.

Johnk, K.D. et al. (2008), Summer heatwaves promote blooms of harmful cyanobacteria, *Global Change Biology*, Vol. 14 (3), pp. 495–512.

Jones, J.A.A. (2009), "Threats to Global Water Security: Population Growth, Terrorism, Climate Change, or Commercialisation?", in: *Threats to Global Water Security, Jones*, J.A.A., T.G. Vardanian and C. Hakopian (Eds.), NATO Science for Peace and Security Series C: Environmental Security.

Joyce, J and Convery, F (2009), Understanding the economics of the Water Framework Directive, *Biology and Environment: Proceedings of the Royal Irish Academy* 109B, 377–83.

Kadlec, R.H., and K.R. Reddy (2001), Temperature effects in treatment wetlands, *Water Environment Research*, Vol. 73, No. 5, pp. 543-557.

Lennartz, S.T. et al. (2014), Long-term trends at the Boknis Eck time series station (Baltic Sea), 1957–2013: does climate change counteract the decline in eutrophication?, *Biogeosciences*, Vol. 11, pp. 6323–6339.

Lieb, C.M. (2004), The Environmental Kuznets Curve and Flow versus Stock Pollution: The Neglect of Future Damages, *Environmental and Resource Economics*, Vol. 29, pp. 483–506.

Ling, N. (2010), Socio-economic drivers of freshwater fish declines in a changing climate: A New Zealand perspective, *Journal of Fish Biology*, Vol. 77, Issue 8, pp. 1983-1992.

Liu, C., N. Hofstra and E. Franz (2013), "Impacts of climate change on the microbial safety of pre-harvest leafy green vegetables as indicated by Escherichia coli O157 and Salmonella spp", *International Journal of Food Microbiology*, Vol. 163(2-3), pp. 119-128.

MfE and StatsNZ (2015), New Zealand's Environmental Reporting Series: Environment Aotearoa 2015. Ministry for the Environment, Wellington, New Zealand.

MfE (2014a), National Policy Statement for Freshwater Management 2014, Ministry for the Environment, Wellington, New Zealand.

MfE (2014b), Delivering Freshwater Reform: A high level overview, Ministry for the Environment, Wellington, New Zealand.

MfE (2013), Freshwater reform 2013 and beyond, Ministry for the Environment, Wellington, New Zealand.

Michalak, A.M. et al. (2013), Record-setting algal bloom in Lake Erie caused by agricultural and meteorological trends consistent with expected future conditions, *Proceedings of the National Academy of Sciences*, Vol. 110 (16), pp. 6448–6452.

Nilsson, C., and B. Malm Renöfält (2008), Linking flow regime and water quality in rivers: a challenge to adaptive catchment management, Ecology and Society Vol. 13, No. 2.

OECD (2015a), *OECD Economic Surveys: New Zealand 2015*, OECD Publishing, Paris, http://dx.doi.org/10.1787/eco_surveys-nzl-2015-en.

OECD (2015b), *Water and Cities: Ensuring Sustainable Futures*, OECD Publishing, Paris, http://dx.doi.org/10.1787/9789264230149-en.

OECD (2014), *Green Growth Indicators 2014, OECD Green Growth Studies*, OECD Publishing, Paris, http://dx.doi.org/10.1787/9789264202030-en.

OECD (2013a), *OECD Compendium of Agri-environmental Indicators*, OECD Publishing, Paris. http://dx.doi.org/10.1787/9789264186217-en.

OECD (2013b), *Water and Climate Change Adaptation: Policies to Navigate Uncharted Waters*, OECD Studies on Water, OECD Publishing, Paris, http://dx.doi.org/10.1787/9789264200449-en.

OECD (2012a), *Water Quality and Agriculture: Meeting the Policy Challenge*, OECD Studies on Water, OECD Publishing, Paris, http://dx.doi.org/10.1787/9789264168060-en.

OECD (2012b), *OECD Environmental Outlook to 2050: The consequences of inaction*, OECD Publishing. http://dx.doi.org/10.1787/9789264122246-en.

OECD (2011), *Towards Green Growth*, OECD Publishing, Paris, http://dx.doi.org/10.1787/9789264111318-en.

Pingali, P.L. (2012) Green Revolution: Impacts, limits, and the path ahead, *Proceedings of the National Academy of Sciences*, Vol. 109, pp. 12302-12308.

Plósz, B.G., Liltved, H. and H. Ratnaweera (2009), Climate change impacts on activated sludge wastewater treatment: a case study from Norway, *Water Science & Technology*, Vol. 60.2, pp. 533-541.

Qin, B.Q. et al. (2010), A drinking water crisis in Lake Taihu, China: linkage to climatic variability and lake management, *Journal of Environmental Management*, Vol. 45 (1), pp. 105–112.

Rabitsch, W. et al. (2016), Developing and testing alien species indicators for Europe, Journal for Nature Conservation, Vol. 29, pp. 89–96.

Ramsar (2015), Wetlands: a global disappearing act, Fact Sheet 3, February 2015, www.ramsar.org/sites/default/files/documents/library/factsheet3_global_disappearing_act_0.pdf.

Royal Commission on Environmental Pollution (2007), *The Urban Environment, Presented to Parliament by Command of Her Majesty March 2007, Westminster, London*. www.gov.uk/government/uploads/system/uploads/attachment_data/file/228911/7009.pdf.

Royal Geographical Society (2012), Water policy in the UK: The challenges. London, UK, RGS-IBG Policy Briefing.

Sadoff, C.W. et al. (2015), Securing Water, Sustaining Growth: Report of the GWP/OECD Task Force on Water Security and Sustainable Growth, University of Oxford, UK, 180pp.

Scavia, D. et al. (2014), Assessing and addressing the re-eutrophication of Lake Erie: Central basin hypoxia, *Journal of Great Lakes Research*, Vol. 40, pp. 226–246.

Sitoki, L., R. Kurmayer, and E. Rott (2012), Spatial variation of phytoplankton composition, biovolume, and resulting microcystin concentrations in the Nyanza Gulf (Lake Victoria, Kenya), *Hydrobiologia*, Vol. 691 (1), pp. 109–122.

Skogen, M.D. et al. (2014), Eutrophication status of the North Sea, Skagerrak, Kattegat and the Baltic Sea in present and future climates: A model study, *Journal of Marine Systems*, Vol. 132, pp. 174–184.

Smith, H.G. et al. (2011), Wildfire effects on water quality in forest catchments: a review with implications for water supply, *Journal of Hydrology*, Vol. 396(1-2), pp. 170-192.

Strayer, D. L. and D. Dudgeon, (2010), Freshwater biodiversity conservation: recent progress and future challenges, *Journal of the North American Benthological Society*, Vol. 29, pp. 344–358.

Thames Tideway Tunnel (2015), https://www.tideway.london/the-tunnel/ [accessed 01/09/2015].

The Guardian (2016), Flint's water crisis: what went wrong. www.theguardian.com/environment/2016/jan/16/flints-water-crisis-what-went-wrong [accessed 29/03/2016].

The Guardian (2015), Michigan city to change water source after studies showed lead increase. www.theguardian.com/us-news/2015/oct/08/flint-michigan-to-change-water-source-lead [accessed 29/03/2016].

UK Environment Agency (2015), Water quality and Water Framework Directive (WFD) classification results: 2014 Progress update. www.gov.uk/government/uploads/system/uploads/attachment_data/file/419090/WQ_trends_2009-2014_March2015.pdf, [accessed online 9 July 2013].

USA Today (2016), How water crisis in Flint, Mich., became federal state of emergency. www.usatoday.com/story/news/nation-now/2016/01/19/michigan-flint-water-contamination/78996052/ [accessed 29/03/2016].

Van Den Berg, J.J. (1993), Effects of sewage sludge disposal, *Land Degradation & Rehabilitation*, Vol. 4, pp. 407-413.

Varjopuro, R. et al. (2014), Coping with persistent environmental problems: systemic delays in reducing eutrophication of the Baltic Sea, *Ecology and Society*, Vol. 19 (4), pp. 48.

Vörösmarty, C.J. et al. (2010), Global threats to human water security and river biodiversity. *Nature*, Vol. 467, No. 7315, pp. 555–561.

Wagenschein, D. and M. Rode (2008), Modelling the impact of river morphology on nitrogen retention-A case study of the Weisse Elster River (Germany), *Ecological Modelling*, Vol. 211, pp. 224-232.

Walther, G.-R. et al. (2009), Alien species in a warmer world: risks and opportunities, *Trends in Ecology and Evolution*, Vol. 24, pp. 686–693.

Walton, B. (2016), Flint Scandal Is Already Changing the Water Utility Business, Circle of Blue. http://www.circleofblue.org/2016/water-policy-politics/flint-scandal-is-already-changing-the-water-utility-business/ [accessed 29/03/2016].

Water UK (2013), CAP Reform: A future for farming and water, Water UK, London, United Kingdom.

Whitehead, P.G. et al. (2009), A review of the potential impacts of climate change on surface water quality, *Hydrological Sciences Journal*, Vol. 54(1), pp. 101-123.

WHO/UNICEF Joint Monitoring Program (JMP) (2015), Progress on Sanitation and Drinking Water: 2015 update and MDG assessment, WHO/UNICEF, Geneva.

World Bank (2015), Introduction to Wastewater Treatment Processes, http://water.worldbank.org/shw-resource-guide/infrastructure/menu-technical-options/wastewater-treatment [accessed 23/07/2015].

WWF (2016), *Living Planet Report 2016: Risk and resilience in a new era*, WWF International, Gland, Switzerland.

WWF (2015), Threat of eutrophication to the Baltic Ecoregion: A widespread and persistent problem, http://wwf.panda.org/what_we_do/where_we_work/baltic/threats/eutrophication/ [accessed 23/07/2015].

Zwolsman, J.J.G. et al. (2011), Water for utilities: Climate change impacts on water quality and water availability for utilities in Europe, European Commission, Technical Report No. 55, WATCH deliverable 6.3.4, pp. 49.

Diffuse Pollution, Degraded Waters
Emerging Policy Solutions
© OECD 2017

Annex 1.A1: An overview of the main water pollutants in OECD countries

Excess nutrient losses

Globally, the most prevalent water quality challenge is eutrophication (UNESCO, 2009), a result of excess nutrient losses (primarily nitrogen and phosphorus) largely from intensive agriculture, and to a lesser extent from urban run-off and wastewater treatment plants. Furthermore, high nutrient loading not only causes eutrophication of lakes and rivers, but also excessive growth of aquatic weeds, algal blooms, hypoxia and declines in the ecosystem functioning of both freshwater and ocean environments.

There are health impacts associated with increasing nitrate toxicity in surface and groundwaters. Nitrate toxicity of drinking water can cause methemoglobinemia in infants (blue-baby syndrome) and other illnesses such as cancer[1]. Nitrates can accumulate in groundwater, particularly in confined aquifers and aquifers with very low recharge rates. It is costly for municipal utilities to invest in additional treatment of drinking water to remove nitrates. The health risk is particularly high in rural areas where untreated contaminated groundwater is abstracted for drinking water from private wells.

Harmful algal blooms are recognised as a global public health threat (Otten and Paerl, 2015). Algal blooms (cyanobacteria, green algae, diatoms) produce toxins which contaminate drinking water, cause acute poisoning, skin irritation and gastrointestinal illness, trigger livestock, fish and shellfish poisoning, and are costly to treat. Their toxicity, unattractive appearance and odour can cause lakeside or riverside property values to decline, and decrease the recreational use and aesthetical value of waters. Algal blooms can also block sunlight and reduce dissolved oxygen to levels that are lethal to aquatic fauna and flora, negatively impacting the food web and inducing fish kills. Algal blooms generally occur in nutrient-enriched waters coupled with low flow, and warm and sunny conditions. They are therefore likely to appear more frequently, and be present for longer, due to higher water temperatures and longer periods of low flows associated with climate change (Bates et al., 2008).

The extent and effects of high nutrient loading extends beyond the freshwater environment, to the development of eutrophic and hypoxic zones (also known as "dead zones") in the oceans. Hypoxic zones have developed in continental seas, such as the Baltic, Kattegat, Black Sea, Gulf of Mexico, and the East China Sea, all of which are major fishery areas (Diaz and Rosenberg, 2008; Robertson and Vitousek, 2009). In total, more than 400 hypoxic dead zones have been identified, and their frequency has approximately doubled each decade since the 1960s (Diaz and Rosenberg, 2008; Robertson and Vitousek, 2009). This trend may continue with the occurrence, frequency, duration and extent of oxygen depletion and harmful algal blooms in coastal zones projected to increase as rivers discharge growing amounts of nutrients into the sea (OECD, 2012a).

A final consideration of nutrient loading in water bodies is the contribution from fallout of air pollutants: nitrogen oxides, sourced primarily from energy and transport emissions, and ammonia sourced primarily from agriculture (in particular, volatilisation of stock manure and effluent).

Microbial contamination

Microbial contamination (pathogenic bacteria and viruses) of water resources is a serious health risk, and is considered the most important pollutant affecting human health globally (Domingo et al., 2007; WWAP, 2009). Water bodies contaminated with pathogens are responsible for the spread of many contagious water-borne diseases (Chapra, 1997), such as cholera, giardiasis and other intestinal infections. This is particularly of concern where contaminated water resources are used for municipal water supply, crop irrigation, and for recreational purposes (Amirat et al., 2012; UK Environment Agency, 2003). Furthermore, microbial contaminants from wastewater can exacerbate biodiversity loss (European Commission, 2013).

Sources of pathogenic microorganisms originate from land application of animal manure, liquid slurry and human biosolids, direct defecation by livestock, leaking septic tanks, and discharges from insufficient wastewater treatment plants. Despite the investment in, and establishment of, wastewater treatment plants, many developed nations still suffer from microbial contamination due to sewage spills and diffuse agricultural pollution, particularly during high rainfall events when combined sewer overflows operate more frequently and runoff from agriculture is induced.

Outbreaks of water-related diseases from contaminated public water supply are not just limited to developing countries. While considerable advances in water supply and sanitation have occurred in OECD countries, pathogens can still enter water supply systems due to resistance to disinfection (e.g. *Giardia, Cryptosporidium*, and enteric viruses), treatment system deficiencies (e.g. inadequate disinfection), periodic treatment failures, or distribution system contamination.

Cases of waterborne disease outbreaks include the Walkerton outbreak of Ontario, Canada, in 2000, which resulted in seven fatalities and hundreds that fell ill as a result of *Escherichia coli O157:H7* infection from contaminated public water supply (Auld et al., 2004). Excessive rainfall and agriculture runoff was the cause of the contamination. In Europe, significant outbreaks of water-borne pathogenic organisms have occurred in Sweden (Widerström et al., 2014; Rehn et al., 2015), Norway (Robertson et al., 2009), Finland (Laine et al., 2011), and throughout England and Wales (Furtado et al., 1998; Smith et al., 2006; Nichols et al., 2009). In New Zealand, the town of Havelock North in the farming region of Hawke's Bay, suffered from a large-scale campylobacter outbreak as recently as August 2016. Over 5 000 people became sick – more than one-third of the town's 14 000 population, causing temporary closure of schools and businesses.

There is also growing evidence of microbial contamination of groundwater, which is often used as an untreated private water supply by many communities, particularly in rural regions (Kay et al., 2007; Feighery et al., 2013). For example, in 2014, 13% of 6 200 private water supplies surveyed in England tested positive *E. coli* (DWI, 2015).

Acidification

Acidification of surface water is particularly acute where there is a strong source of acid, such as downwind of atmospheric pollutants (acid deposition), particularly where soils have a low buffering capacity, and downstream of mining areas (acid mine drainage). A reduction in sulphur pollutants through control measures at power plants has reduced the impact and occurrence of acid deposition. However, nitrogen oxides (NO_x) (largely from combustion processes thermal and fossil fuel), and ammonia (NH_3) (largely from livestock manure and urine, synthetic nitrogen fertilisers and the cultivation of legumes and other crops), are still prominent sources of atmospheric pollutants that cause acid deposition, eutrophication, contribute indirectly to climate change, and reduce visibility (acid aerosols/smog).

Acid mine drainage (AMD) (also known as acid rock drainage) is often considered one of the main pollutants of water in countries that have historic or current mining activities (Simate and Ndlovu, 2014). AMD is a strong acid wastewater, rich in high concentrations of dissolved heavy metals. It most commonly originates from operating and abandoned polymetallic sulphide mining sites, where exposure of sulphide minerals to oxygen, water and microorganisms triggers chemical, biological, and electrochemical reactions that promote the creation of sulphuric acid, which subsequently promotes the release of heavy metals (Egiebor and Oni, 2007; Jennings et al., 2008; Simate and Ndlovu, 2014).

The acidification and mobilisation of heavy metals in water bodies has severe impacts on human and aquatic species health, can contaminate groundwater, raise water treatment costs, impact commercial fisheries, and cause corrosion of concrete and metal infrastructure. Impacts to aquatic species include fish kill, stunted growth, reduced reproduction, deformities, and disruption of the aquatic food chain (Jennings et al., 2008). Impacts to human health include disruption of metabolic functions; liver, kidney, and gastrointestinal damage; damage to the nervous system; and other health effects (Simate and Ndlovu, 2014).

Despite efforts to prevent, mitigate and control AMD using the best available technologies, AMD production can be sustained for hundreds of years and remains the greatest environmental liability associated with mining (Egiebor and Oni, 2007; Jennings et al., 2008).

Salinity

Salinity is the primary threshold that limits water use and availability. Groundwater quality deterioration is of concern particularly in coastal regions, and in semi-arid regions with salinised soils. Clearing of perennial vegetation and irrigation of salt-affected soils leads to leaching of salts and increasing groundwater salinity (OECD, 2012b). In addition, rising water tables of high salinity can lead to salinisation of soils and cause a positive feedback mechanism for further increasing salinity of groundwater. In coastal areas, over-abstraction of groundwater, and sea level rise associated with climate change, contributes to salt water intrusion into aquifers. The application of salt for anti-icing and de-icing of roads can also enter surface water bodies. The principle processes causing groundwater salinity are illustrated in Figure A1 below.

Groundwater salinity and soil salinisation are largely irreversible and have impacts on irrigation, and domestic and industry water use, as well as long term effects on agricultural land and aquatic life (Bennett et al., 2009). The effects, and the continued considerable threat, of salt water intrusion on the global scale is well documented (e.g. OECD, 2015; Werner et al., 2013). A number of OECD countries are especially concerned with seawater intrusion to aquifers in coastal areas, such as Greece, Italy, Spain, the Netherlands, Australia, New Zealand, Mexico and parts of the United States (OECD, 2015).

Figure A1. **The principle processes causing groundwater salinisation**

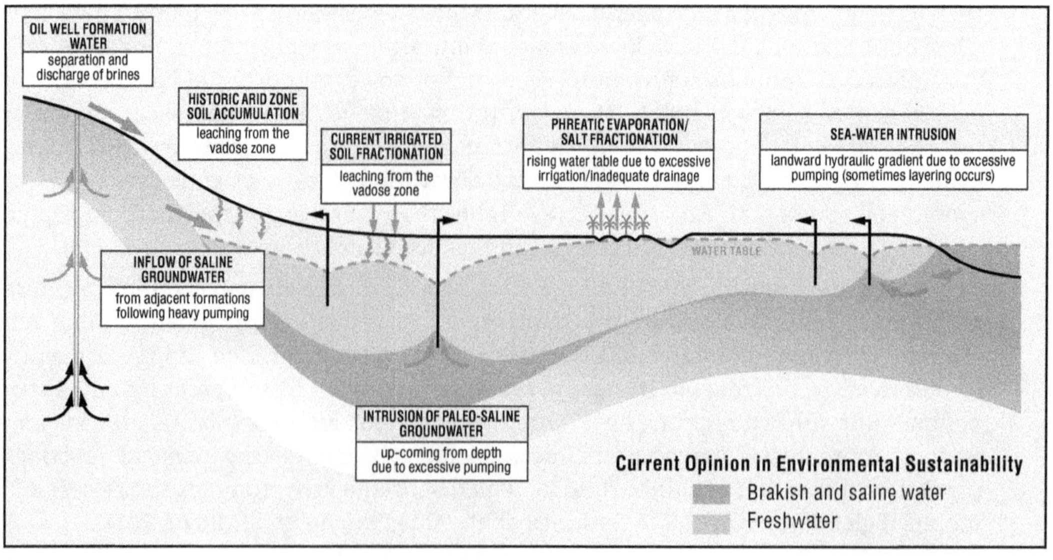

Source: Foster et al. (2013).

Sedimentation and organic materials

Sedimentation and the release of organic matter (dissolved organic carbon [DOC]), particularly during storm events and sediment release from dams, can cause turbidity and deoxygenation of water bodies. Deoxygenation of water causes loss of aquatic life. Sedimentation and turbidity blocks light and inhibits aquatic plant growth, reduces die-off of pathogens, clogs the gills of fish and smothers benthic invertebrates and fish eggs. It effects water transportation systems, prevents effective treatment of drinking water and wastewater, and produces carcinogenic compounds (such as chloroform) when water with high levels of DOC is disinfected with chlorine.

Sediment particles can also carry heavy metals, pathogens, pesticides and other toxic organic compounds as attachments. In the United Kingdom, DOC concentrations in 22 upland catchments increased by an average of 91% between 1988 and 2003 (Evans et al., 2004). Here, the increase in DOC has significantly affected other chemical variables, in particular, an increase in metal transport (organic Al and Fe). In England and Wales, past and present mining, processing and utilisation of base metals have polluted sediments with heavy metals in rivers, lakes and estuaries. Such sediments are likely to be causing ecological damage; the re-suspension of these sediments during floods has the potential to cause additional harm to aquatic life and contaminate floodplain soils used for agriculture, and this has implications for delivering the Water Framework Directive (UK Environment Agency, 2008; DEFRA, 2003). Increases in DOC have also occurred elsewhere in Europe and North America (Evans et al., 2004).

Human-induced sources of DOC and sediment in surface water bodies include food processing waste, manure spreading on livestock farms, sediment release from dams, and erosion of topsoil and peatlands due to poor farming, construction, and forestry (including deforestation) practices (Neal and Hill, 1994). Dams retain sediment the majority of the time and regulate flow regimes. This alters the timing and intensity of river discharge, causing downstream riverbed and delta erosion and affecting the natural imbalance of sediment, nutrients and organic matter to flood plains used for agricultural production and coastal marine ecosystems (e.g. Renshaw et al., 2014; Yang et al., 2014). Such imbalances can also increase the risk of harmful algal blooms (OECD, 2012a).

Toxic contaminants

Toxic contaminants include pesticides, heavy metals, chlorinated solvents, and persistent organic pollutants (POPs).

Pesticides (which include herbicides, insecticides, fungicides and other chemical agents) are introduced to surface and groundwater from their use for plant and animal protection in agriculture. In nearly half of OECD countries, nutrient and pesticide concentrations in surface and groundwater in agricultural areas exceed national recommended limits for drinking water standards (OECD, 2012a). Furthermore, pest species can become resistant to the harmful effects of pesticides over time through genetic adaptation (Becker and Liess, 2015). For example, genetic resistance to herbicides has been recorded in 210 weed species (Bourguet et al., 2013). Similarly, genetic resistance has been recorded against 300 insecticide compounds in over 500 pest species, and 30 fungicides in 250 species of phytopathogenic fungi. This adaptive evolution reduces the efficacy of conventional control strategies and requires the development of new pesticides (which is costly and time consuming) in order to keep up with increasing resistance. However, interspecific competition and predation can delay genetic adaptation to pesticides, supporting the importance of biodiversity for effective pest management (Becker and Liess, 2015). Other factors that can potentially delay the onset of pesticide resistance include the deployment of biopesticides (e.g. bio-based products); implementation of integrated pest management (IPM) programme; use of genetically engineered crops; adoption of sensitivity monitoring for pesticide resistance; and use of pest forecasting advisory systems.

Land application of sewage sludge, urban stormwater runoff, corroding water supply infrastructure, and industrial waste are sources of heavy metal pollution to water resources. Furthermore, soil and water acidification (such as acid mine drainage) and flooding events can exacerbate these issues, mobilising heavy metal contaminants in sediment and rock. Some regions of the world suffer from naturally elevated levels of heavy metals (such as arsenic and chromium) in groundwater due to the local geology. Natural arsenic pollution of drinking water is now considered a global threat with as many as 140 million people affected in 70 countries on all continents (UNESCO, 2009). For example, well-known areas with high arsenic concentrations include Argentina, Chile, Mexico, China and Hungary, and more recently in West Bengal (India), Bangladesh and Vietnam. In such areas, more than 10% of wells may be "affected" (defined as those exceeding 50 µg/L) and in the worst cases, this figure may exceed 90% (Smedley and Kinniburgh, 2002).

The contamination of water bodies with chlorinated solvents (e.g. petrol and aviation fuel) originates from transport, industrial spills, leaking underground storage tanks, hydraulic fracking, and urban stormwater runoff. For example, hydraulic fracturing involves the injection of fluids under pressures great enough to fracture oil- and gas-producing formations (U.S. EPA, 2015a). The fluid generally consists of water, chemicals, and proppant (commonly sand). Concerns have risen regarding not only competition for the large quantities of water used, but also regarding groundwater and surface water quality (Ground Water Protection Council, 2009; Verrastro and Branch, 2010; Grubert and Kitase, 2010; Center for Biological Diversity, 2014).

Three potential pathways for water contamination have been identified through the hydraulic fracking process: (1) the contamination of shallow aquifers with fugitive hydrocarbon gases (i.e., stray gas contamination), which can also potentially lead to the salinisation of shallow groundwater through leaking natural gas wells and subsurface flow; (2) the contamination of surface water and shallow groundwater from spills, leaks, and/or the disposal of inadequately treated wastewater; and (3) the accumulation of toxic and radioactive elements in soil or stream sediments near disposal or spill sites (Rozell and Reaven, 2012; Vengosh et al., 2014; Vengosh et al., 2013; Vidic et al., 2013). In the United States, oil and gas production via hydraulic fracturing in shales, tight formations and coalbeds occurs in close proximity (within 1 mile) of approximately 6 800 public water supply sources (U.S. EPA, 2015a). Although the number of identified cases where drinking water resources have been impacted are small relative to the number of hydraulically fractured wells, the risk is significant with such public water supply sources serving more than 8.6 million people. In recognition of the risks of water pollution from hydraulic fracturing, the United States EPA is improving the scientific understanding of contamination and its effects, and providing guidance for its management (U.S. EPA, 2015b).

Persistent organic pollutants (POPs) are organic chemical substances that are resistant to biodegradation and have been linked to bioaccumulation in the food web, declines in a number of bird species, and poor human and environmental health (U.S. EPA, 2009). The use of chemicals and synthetic pesticides with POPs has largely been phased out in developed countries due to the signing of the Stockholm Convention 2001, in which countries agreed to reduce or eliminate the production, use, and/or release of 12 key POPs (e.g. DDT, dioxins, and PCBs[2]). However, due to their long life-time, the ban on POPs may not necessarily lead to a sharp reduction in their occurrence (Lohmann et al., 2007).

Thermal pollution

Thermal pollution results in a change in the physical properties of water. Causes of thermal pollution are most commonly associated with power plants and industrial manufacturers who use water as a coolant before discharging it back to the environment. Urban runoff may also elevate the temperature of surface waters. Elevated water temperatures decrease oxygen levels, which can kill fish, alter food chain composition, reduce species biodiversity, and foster invasion by new thermophilic species (e.g. Teixeira et al., 2009; Chuang et al., 2009). Thermal pollution can also be caused by the release of very cold water from the base of reservoirs into warmer rivers (Langford, 2001). The global extent of thermal pollution is not well-documented, and limited research has been undertaken on its impact to freshwater ecosystems (most are marine studies).

Plastic particle pollution

Pollution of water bodies with plastic (micro-plastic and macro-plastic) and other solid waste is a significant problem, caused by illegal and accidental rubbish dumping by individuals, the plastic production industry, recreational and commercial fishers, and urban stormwater runoff. Beaches and oceans act as a sink, and since plastic is also buoyant, an increasing load of plastic debris is being dispersed over long distances. When the debris does finally settle in sediments, it may persist for centuries (Derraik, 2002).

Plastic particle pollution disrupts the marine food chain, causes physiological damage to marine organisms through ingestion and entanglement, and releases POPs (such as PCBs) into the surrounding waters. Plastic particle pollution is an increasing problem, and it is reported that 88% of reported incidents between organisms and total marine debris was associated with plastic, 11% of which was with microplastics (GEF, 2012). For instance, surveys carried out in the western North Atlantic Ocean show significant increases in plastic particle density from 1991 to 2007 (Morét-Ferguson et al., 2010). A series of surveys in coastal Australia, Bay of Bengal and the Mediterranean Sea estimates that a minimum of 5.25 trillion plastic particles weighing over 260 000 tons are afloat at sea in the world's oceans (Eriksen et al., 2014). In the Mediterranean Sea, plastic debris has been found in the stomachs of 18% of swordfish, bluefin tuna and albacore (Romeo et al., 2015). Due to their small size, microplastics may be ingested by low trophic fauna, with uncertain consequences for the health of organisms (Wright et al., 2013). There is concern that the POPs in plastic can have additional toxic effects on fish (such as reproductive changes) and impacts on human health from the transfer of these chemicals via consumption of fish. Further research is required regarding these ecological and human health effects.

Contaminants of emerging concern

Contaminants of emerging concern (CECs) comprise a vast array of contaminants that have only recently appeared in water, or that are of recent concern because they have been detected at concentrations significantly higher than expected, and/or their risk to human and environmental health may not be fully understood. Examples include pharmaceuticals, hormones, industrial chemicals, personal care products, perfluorinated compounds, flame retardants, plasticizers, detergent compounds, caffeine, fragrances, cyanotoxins, engineered nanomaterials (such as carbon nanotubes or nano-scale particulate titanium dioxide), anti-microbial cleaning agents and their transformation products[3]. The number of CECs is continuously evolving as new chemical compounds are produced, and improvements in science and monitoring increase our understanding of the effects of current and past contaminants on human and environmental health (Sauvé and Desrosiers, 2014). Despite the increasing number of published studies covering CECs, very little is known about the effects of their transformation products and/or metabolites (Lambropoulou and Nollet, 2014).

CECs are derived from a variety of municipal, domestic, agricultural, and industrial waste sources. Complex mixtures of CECs have been documented in plant and animal tissue, groundwater and drinking water (Khan et al., 2016; Pal et al., 2014). There is also growing concern that CECs may be bioactive and interactive (e.g. with additive, synergistic and/or antagonistic effects) (Bhandari et al., 2009). Large CEC concentrations have been identified in a variety of sources such as septic systems, landfills, animal manure and sewage sludge (Pal et al., 2014).

Endocrine disruptors affect the reproductive systems of fish and humans, with some documented cases of fish and other freshwater and marine organisms changing sex (Porte et al., 2006; Velasco-Santamaría et al., 2011; Vos et al., 2000). But perhaps the CEC of most worry is excess use of antibiotics and the health implications of increasing resistance to antibiotics.

Antibiotic resistance is now considered by authorities in the United States, United Kingdom and by international experts to be one of the paramount public health challenges of our time (Centers for Disease Control and Prevention, 2013; UK Department of Health, 2013; WHO, 2014). Antibiotics kill good bacteria in rivers which naturally treat pollution, and build human resistance to antibiotics.

Antibiotics residues, and antibiotic resistant bacteria and genes, from the intentional use of antibiotics (for human and animal health) can enter the environment (water and soil) through wastewater systems (Proia et al., 2016), direct livestock excretion, and land application of biosolids and effluent (Williams-Nguyen et al., 2016). They then have the potential to impact on human health, ecosystem functioning and agricultural productivity (although research on the pathways of exposure is premature). For example, a study by Khan et al. (2016) discovered the occurrence of antibiotic resistant bacteria exist at the consumers' end of the drinking water distribution system in Glasgow, Scotland, some of which also contain integrase genes, which can aid in the dispersion of resistance genes.

The current additional economic costs of antimicrobial resistance is considered to be relatively small, as mitigating action (switching prescribing to other antibiotics) can be taken. However, when this is coupled with growing rates of resistant organisms and a declining number of new antibiotics in development, the potential future health costs are very high (Smith and Coast, 2012).

Other factors that contribute to degradation of water quality

Other factors that contribute to degradation of freshwater ecosystems, and thus their ability to process contaminants, include the introduction of invasive alien species and anthropogenic modifications to river systems. According to the IUCN, invasive alien species constitute the second most severe threat to freshwater fish species (William et al., 2009), and the spread of invasive alien species is projected to increase due to a combination of increasing trade and climate change (Death et al., 2015; Rabitsch et al., 2016; Walther et al., 2009). Changes in the natural morphology of water bodies (e.g. channelized rivers, dams, canals) can also affect the ability of ecosystems to process and retain pollutants (Wagenschein and Rode, 2008).

The deterioration of water quality also has subsequent knock-on impacts on the functioning of in-stream invertebrates, fish, and aquatic plant communities (Doledec et al., 2006; Ling, 2010), which causes negative feedbacks, particularly the ability of ecosystems to process contaminants, thereby causing pollutants to accumulate in the environment and cause further damage.

Notes

1. High nitrate levels in drinking water can lead to infant methaemoglobinaemia (blue-baby syndrome), gastric cancer goiter, metabolic disorder, birth malformations, hypertension and livestock poisoning (Khandare, 2013). This is of particular concern in private groundwater wells located in agricultural regions with overlying permeable soils. The US Environmental Protection Agency requires that public water systems not exceed a nitrate concentration of 10 mg/L, because nitrate above this level, causes infant methaemoglobinaemia.
2. DDT: dichlorodiphenyltrichloroethane, a synthesised organochloride known for its insecticidal properties; PCBs: Polychlorinated biphenyls, a synthetic organic chemical compound that was widely used as electrical insulators, coolant fluid, flame retardants, and plasticisers in paints and cements, amongst other uses.
3. Transformation products and metabolites of man-made chemicals that are produced from biological, chemical and physical breakdown reactions.

References

Amirat, L. et al. (2012) "Escherichia coli contamination of the river Thames in different seasons and weather conditions", *Water and Environment Journal*, Vol. 26, No. 4, pp. 482-489.

Auld, H., MacIver, D. and J. Klaassen (2004), "Heavy rainfall and waterborne disease outbreaks: The Walkerton example", *Journal of Toxicology and Environmental Health-Part a-Current Issues*, Vol. 67, No. 20-22, pp. 1879-1887.

Bates, B.C., Z.W. Kundzewicz, S. Wu and J.P. Palutikof, Eds. (2008), Climate Change and Water. Technical Paper of the Intergovernmental Panel on Climate Change, IPCC Secretariat, Geneva, 210 pp.

Becker, J.M. and M. Liess (2015), Biotic interactions govern genetic adaptation to toxicants. Proceedings of the Royal Society B, Vol. 282: 20150071.

Bennett, S.J., E.G. Barrett-Lennard, T.D. Colmer (2009), Salinity and waterlogging as constraints to saltland pasture production: a review, *Agriculture, Ecosystems and Environment*, Vol. 123, pp. 349–360.

Bhandari, A. et al. (2009), Contaminants of Emerging Environmental Concern. American Society of Civil Engineers, Reston, Virginia, pp.492.

Bourguet, D. et al. (2013), Heterogeneity of selection and the evolution of resistance, prepared by the Resistance to Xenobiotics (REX) Consortium, *Trends Ecology & Evolution*, Vol. 28, pp. 110–118.

Center for Biological Diversity (2014), "Documents Reveal Billions of Gallons of Oil Industry Wastewater Illegally Injected Into Central California Aquifers: Tests Find Elevated Arsenic, Thallium Levels in Nearby Water Wells", Press Release October 6, 2014, Center for Biological Diversity, Tucson, Arizona, USA.

Centers for Disease Control and Prevention (2013), Antibiotic resistance threats in the United States, 2013. US Department of Health and Human Services, Washington, DC.

Chapra, S. C., (1997) *Chapter 27. Pathogens. In: Surface Water-Quality Modelling*, USA: McGraw Companies.

Chuang, Y., Yang, H. and H. Lin (2009), Effects of thermal discharge from a nuclear power plant on phytoplankton and periphyton in subtropical coastal waters, *Journal of Sea Research*, Vol. 61, pp. 197-205.

Death, R.G., Fuller, I.C. and M.G. Macklin (2015), Resetting the river template: the potential for climate-related extreme floods to transform river geomorphology and ecology, *Freshwater Biology*, Vol. 60, pp. 2477–2496.

DEFRA (2003), *The use of geomorphological mapping and modelling for identifying land affected by metal contamination on river floodplains*, Department for Environment, Food and Rural Affairs, London, UK.

Derraik, J.G.B (2002), "The pollution of the marine environment by plastic debris: a review", *Marine Pollution Bulletin*, Vol. 44, No. 9, pp. 842-852.

Diaz, R.J. and R. Rosenberg (2008), "Spreading dead zones and consequences for marine ecosystems", *Science*, Vol. 321, No. 5891, pp. 926-929.

Doledec, S. et al. (2006) 'Comparison of structural and functional approaches to determining landuse effects on grassland stream invertebrate communities', Journal of the North American Benthological Society, 25(1), 44-60.

Domingo, J. W. S. et al. (2007) "Quo vadis source tracking? Towards a strategic framework for environmental monitoring of fecal pollution", *Water Research*, Vol. 41, No. 16, pp. 3539-3552.

DWI (2015), *Drinking water Drinking water quality in England: The position after 25 years of regulation*. July 2015. A report by the Chief Inspector of Drinking Water, Drinking Water Inspectorate, London.

Egiebor, N.O. and B. Oni (2007), Acid rock drainage formation and treatment: a review, *Asia-Pacific Journal of Chemical Engineering, Vol. 2, pp. 47-62.*

Eriksen, M. et al. (2014), Plastic Pollution in the World's Oceans: More than 5 Trillion Plastic Pieces Weighing over 250,000 Tons Afloat at Sea. *PLoS ONE*, Vol. 9, No. 12, pp. 1-15.

European Commission (2013), Seventh Report on the Implementation of the Urban Waste Water Treatment Directive (91/271/EEC), European Commission, Brussels, Belgium.

Evans, C.D., Monteith, D.T. and D.M. Cooper (2004), "Long-term increases in surface water dissolved organic carbon: Observations, possible causes and environmental impacts", *Environmental Pollution*, Vol. 137, pp. 55-71.

Feighery, J. et al. (2013), "Transport of E-coli in aquifer sediments of Bangladesh: Implications for widespread microbial contamination of groundwater", *Water Resources Research*, Vol. 49, No. 7, pp. 3897-3911.

Foster, S., J. et al. (2013), "Groundwater — a global focus on the "local resource", *Current Opinion in Environmental Sustainability*, Vol. 5(6), pp. 685–695.

Furtado, C. et al. (1998), "Outbreaks of waterborne infectious intestinal disease in England and Wales, 1992-5", *Epidemiology and Infection*, Vol. 121, No. 1, pp. 109-119.

GEF (2012), Impacts of Marine Debris on Biodiversity: Current Status and Potential Solutions, Technical Series No. 67, Secretariat of the Convention on Biological Diversity and Scientific and Technical Advisory Panel – GEF, Montreal, pp. 61.

Ground Water Protection Council (2009), State Oil and Natural Gas Regulations Designed to Protect Water Resources (Oklahoma City, Okla.: Ground Water Protection Council).

Grubert, E. and S. Kitase (2010), "How Energy Choices Affect Fresh Water Supplies: A Comparison of U.S. Coal and Natural Gas," World Watch Institute.

Jennings, S.R., Neuman, D.R. and P.S. Blicker (2008), *Acid Mine Drainage and Effects on Fish Health and Ecology: A Review. Reclamation Research Group Publication, Bozeman, MT.*

Kay, D. et al. (2007), "The microbiological quality of seven large commercial private water supplies in the United Kingdom", *Journal of Water and Health*, Vol. 5, No. 4, pp. 523-538.

Khan, S; Knapp, C.W. and T.K. Beattie (2016), Antibiotic Resistant Bacteria Found in Municipal Drinking Water, *Environmental Processes*, online March 2016, pp 1-12.

Khandare, H.W. (2013), Scenario of nitrate contamination in groundwater: Its causes and prevention, International Journal of ChemTech Research, Vol. 5, No 4, pp. 1921-1926.

Laine, J. et al. (2011), "An extensive gastroenteritis outbreak after drinking-water contamination by sewage effluent, Finland", *Epidemiology and Infection*, Vol. 139, No. 7, pp. 1105-1113.

Lambropoulou, D.A. and L.M.L. Nollet (2014), Transformation Products of Emerging Contaminants in the Environment: Analysis, Processes, Occurrence, Effects and Risks.

Langford, T.E.L., 2001. Thermal discharges and pollution. In: Thorpe, S.A., Turekian, K.K. (Eds.), Encyclopedia of Ocean Sciences, vol. 6. Elseiver Science Ltd.

Ling, N. (2010), Socio-economic drivers of freshwater fish declines in a changing climate: A New Zealand perspective, Journal of Fish Biology, Vol. 77, Issue 8, pp. 1983-1992.

Lohmann, R. et al. (2007), "Global fate of POPs: Current and future research directions", *Environmental Pollution, Vol. 150, pp. 150-165.*

Morét-Ferguson, S. et al. (2010), "The size, mass, and composition of plastic debris in the western North Atlantic Ocean", *Marine Pollution Bulletin*, Vol. 60, pp. 1873–1878.

Neal, C. and S. Hill (1994), "Dissolved inorganic and organic carbon in moorland and forest streams: Plynlimon, Mid-Wales", *Journal of Hydrology*, Vol. 153, No. 1–4, pp. 231-243.

Nichols, G. et al. (2009), "Rainfall and outbreaks of drinking water related disease and in England and Wales", *Journal of Water and Health*, Vol. 7, No. 1, pp. 1-8.

OECD (2015), Drying wells, rising stakes: towards sustainable agricultural groundwater use, OECD Studies on Water, OECD Publishing, Paris. http://dx.doi.org/10.1787/9789264238701-en.

OECD (2012a), *OECD Environmental Outlook to 2050: The consequences of inaction*, OECD Publishing, Paris, http://dx.doi.org/10.1787/9789264122246-en.

OECD (2012b), *Water Quality and Agriculture: Meeting the Policy Challenge*, OECD Studies on Water, OECD Publishing, Paris, http://dx.doi.org/10.1787/9789264168060-en.

Otten, T.G. and H.W. Paerl (2015), Health Effects of Toxic Cyanobacteria in U.S. Drinking and Recreational Waters: Our Current Understanding and Proposed Direction, Current Environmental Health Reports, Vol. 2, pg 75–84.

Porte C. et al. (2006), Endocrine disruptors in marine organisms: approaches and perspectives, *Comparative Biochemistry and Physiology Part C: Pharmacology, Toxicology and Endocrinology*, Vol. 143, No. 3, pp. 303–315.

Proia, L. et al. (2016), Occurrence and persistence of antibiotic resistance genes in river biofilms after wastewater inputs in small rivers, *Environmental Pollution*, Vol. 210, pp.121–128.

Rabitsch, W. et al. (2016), Developing and testing alien species indicators for Europe, Journal for Nature Conservation, Vol. 29, pp. 89–96.

Rehn, M. et al. (2015), Post-infection symptoms following two large waterborne outbreaks of *Cryptosporidium hominis in Northern Sweden*, 2010–2011, BMC Public Health, Vol. 15: 529.

Renshaw, C.E. et al. (2014), Impact of flow regulation on near-channel floodplain sedimentation, *Geomorphology*, Vol. 205, pp. 120-127.

Robertson, L. et al. (2009), "A water contamination incident in Oslo, Norway during October 2007; a basis for discussion of boil-water notices and the potential for post-treatment contamination of drinking water supplies", *Journal of Water and Health*, Vol. 7, No. 1, pp. 55-66.

Robertson, G.P. and P.M Vitousek (2009), "Nitrogen in Agriculture: Balancing the Cost of an Essential Resource", *Annual Review of Environment and Resources*, Palo Alto, Annual Reviews, pp. 97-125.

Romeo, T. et al. (2015), First evidence of presence of plastic debris in stomach of large pelagic fish in the Mediterranean Sea, *Marine Pollution Bulletin*, Vol. 95, pp. 358–361.

Rozell, D.J. and S.J. Reaven (2012), Water Pollution Risk Associated with Natural Gas Extraction from the Marcellus Shale, *Risk Analysis*, Vol. 32, No. 8.

Sauvé1, S. and M. Desrosiers (2014), A review of what is an emerging contaminant, *Chemistry Central Journal*, Vol. 8, No. 15.

Simate, G.S and S. Ndlovu (2014), Acid mine drainage: Challenges and opportunities, *Journal of Environmental Chemical Engineering*, Vol. 2, pp. 1785–1803.

Smedley, P.L. and D.G. Kinniburgh (2002), A review of the source, behaviour and distribution of arsenic in natural waters (review), *Applied Geochemistry*, Vol. 17, No. 5, pp. 517-568.

Smith, A. et al. (2006), "Outbreaks of waterborne infectious intestinal disease in England and Wales, 1992-2003", *Epidemiology and Infection*, Vol. 134, No. 6, pp. 1141-1149.

Smith R and J. Coast (2012). The economic burden of antimicrobial resistance: Why is it not more substantial? Technical Report. London School of Hygiene Tropical Medicine, London.

Teixeira, T.P., Neves, L.M. and F.G. Araujo (2009), "Effects of a nuclear power plant thermal discharge on habitat complexity and fish community structure in Ilha Grande Bay, Brazil", *Marine Environmental Research*, Vol. 68, pp. 188-195.

UK Department of Health (2013), Annual Report of the Chief Medical Officer, Volume Two, 2011: Infections and the rise of antimicrobial resistance.

UK Environment Agency (2003) *The Microbiology of Sewage Sludge (2003) - Part 1 - An overview of the treatment and use in agriculture of sewage sludge in relation to its impact on the environment and public health*, Environment Agency, UK.

UK Environment Agency (2008), *Assessment of Metal Mining-Contaminated River Sediments in England and Wales*, Science Report: SC030136/SR4, Environment Agency, UK.

UNESCO (2009), *The United Nations World Water Development Report 3: Water in a Changing World*. World Water Assessment Programme, UNESCO, Paris.

U.S. EPA (2015a), *Assessment of the Potential Impacts of Hydraulic Fracturing for Oil and Gas on Drinking Water Resources* (External Review Draft), U.S. Environmental Protection Agency, Washington, DC, EPA/600/R-15/047, 2015.

U.S. EPA (2015b), *Natural Gas Extraction - Hydraulic Fracturing*, U.S. Environmental Protection Agency, Washington, DC, [online] https://www.epa.gov/hydraulicfracturing [accessed 09/09/2015].

U.S. EPA (2009), Persistent Organic Pollutants: A Global Issue, A Global Response, https://www.epa.gov/international-cooperation/persistent-organic-pollutants-global-issue-global-response.

Velasco-Santamaría, Y.M. et al. (2011), Bezafibrate, a lipid-lowering pharmaceutical, as a potential endocrine disruptor in male zebrafish (Danio rerio), *Aquatic Toxicology*, Vol. 105, pp. 107– 118.

Vengosh, A. et al. (2014), A critical review of the risks to water resources from unconventional shale gas development and hydraulic fracturing in the United States, *Environmental Science and Technology*, Vol. 48, pp. 8334–8348.

Vengosh A. et al. (2013), The effects of shale gas exploration and hydraulic fracturing on the quality of water resources in the United States, *Procedia Earth and Planetary Science*, Vol. 7, pp. 863–866.

Verrastro F. and C. Branch (2010), "Developing America's Unconventional Gas Resources: Benefits and Challenges," Center for Strategic and International Studies.

Vidic, R. D. et al. (2013), Impact of Shale Gas Development on Regional Water Quality, *Science*, Vol. 340.

Vos, J.G. et al. (2000), Health effects of endocrine-disrupting chemicals on wildlife, with special reference to the European situation, *Critical Reviews in Toxicology*, Vol. 30, No. 1, pp. 71-133.

Wagenschein, D. and M. Rode (2008), Modelling the impact of river morphology on nitrogen retention-A case study of the Weisse Elster River (Germany), *Ecological Modelling*, Vol. 211, pp. 224-232.

Walther, G.-R. et al. (2009), Alien species in a warmer world: risks and opportunities, *Trends in Ecology and Evolution*, Vol. 24, pp. 686–693.

Werner, A.D. et al. (2013), Seawater intrusion processes, investigation and management: Recent advances and future challenges, *Advances in Water Resources*, Vol. 51, pp. 3–26.

WHO (2014), Antimicrobial resistance: Global report on surveillance 2014, WHO, Geneva, Switzerland.

Widerström, M. et al. (2014), Large Outbreak of *Cryptosporidium hominis Infection Transmitted through the Public Water Supply, Sweden, Emerging Infectious Diseases*, Vol. 20(4), pp. 581–589.

William, R.T. et al. (2009), Freshwater biodiversity: a hidden resource under threat, *In*: Vié, J.-C., Hilton-Taylor, C. and S.N. Stuart (eds.), Wildlife in a Changing World – An Analysis of the 2008 IUCN Red List of Threatened Species, IUCN, Gland, Switzerland 180 pp.

Williams-Nguyen, J. et al. (2016), Antibiotics and Antibiotic Resistance in Agroecosystems: State of the Science, Journal of Environmental Quality, February 2016.

Wright, J. (2013) *Water quality in New Zealand: Land use and nutrient pollution*, November 2013, Wellington: Parliamentary Commissioner for the Environment.

WWAP (World Water Assessment Programme) (2009), The United Nations World Water Development Report 3: Water in a Challenging World. Paris, UNESCO, and London, Earthscan.

Yang, S.L. et al. (2014), Downstream Sedimentary and Geomorphic Impacts of The Three Gorges Dam on the Yangtze River (Review), *Earth-Science Reviews*, Vol. 138, pp. 469-486.

Diffuse Pollution, Degraded Waters
Emerging Policy Solutions
© OECD 2017

Chapter 2

Economic costs and policy approaches to control diffuse source water pollution

This chapter looks at the impacts and costs of water pollution to society and argues who should pay for, and benefit from, improvements in water quality. The chapter lastly inventories the range of policies in place in OECD countries to manage water quality and discusses the importance of policy coherence across policy domains for the management of diffuse pollution.

Key messages

The cost of delaying further improvements in water quality is significant for OECD countries. Despite challenges with economic valuation, national estimates suggest that the cost of water pollution in OECD countries is likely to exceed billions of dollars each year. Such estimates serve to illustrate the existence of significant externalities and a need to adjust water, urban and agriculture management practices in order to reduce negative impacts on water quality.

A number of complex variables determine the impact of pollution on water bodies and therefore influence policy responses to control them. Pollution events are unevenly distributed, both spatially and temporally, and there are ecological and social response time delays that make management of water pollution, particularly diffuse pollution, a complex task.

Markets for agricultural commodities do not internalise water pollution externalities nor signal their value to producers or consumers. This market failure is a difficult policy challenge and one that justifies intervention by government or communities to achieve more economically, environmentally and socially optimal and sustainable outcomes. An important policy area to be examined is the increasing need to find cost-effective solutions and economic instruments that incentivise pollution reduction and fund water quality improvements, particularly given fiscal consolidation of government budgets.

Water quality improvements come at a cost. When developing policy to manage water quality, an important consideration is not only the measurement of the costs and benefits of water pollution reductions, but also on to whom these costs and benefits will fall. The Polluter Pays Principle has typically not been successful in the control of diffuse pollution because of the limitations on measurement, abatement measures, poor enforcement, and political resistance. The Beneficiary Pays Principle has had more success as an incentive to reduce diffuse pollution, particularly on a voluntary basis, but can cause equity issues if polluters are seen to be rewarded.

A lack of policy coherence across agricultural, urban, energy, industrial, economic, climate, environmental and water policies has failed to avoid conflicting signals and incentives to users of water. Special consideration of the cross-sectoral nature and potential trade-offs between climate policies and water quality will need to be managed as climate mitigation and adaptation policies are developed (e.g. bioenergy crops can increase water demand and decrease water quality and food security; afforestation of water catchments reduces soil erosion and local flood risk, and improves water quality).

The economic case for water quality management

Water quality risks

Poor water quality has many economic costs associated with it, including: *i)* degradation of ecosystem services; *ii)* water treatment and health-related costs; *iii)* impacts on economic activities such as agriculture, fisheries, industrial manufacturing and tourism; *iv)* reduced property values; and *v)* opportunity costs of further development (WWAP, 2012). For example, risks associated with deterioration of water quality to agriculture include unusable water resources for irrigation, irreversible groundwater and soil contamination, and health effects on livestock, plants and humans (OECD, 2013). Examples of water quality impacts to economic, social and environmental values are presented in Table 2.1.

Table 2.1. **Impacts of water pollution: Economic, social and environmental**

Impact	Examples
Human health	Polluted water is the world's largest health risk, and continues to threaten both quality of life and public health. Associated with this are health service costs, loss life expectancy, and emergency health costs associated with major pollution events.
Ecosystem health	Damage to freshwater and marine ecosystems (e.g. fish kill, invertebrates, benthic fauna, flora, habitat degradation) and loss of ecosystem services (including the ability to process pollutants), which may require investment in additional or different grey infrastructure alternatives to replicate these services.
Social values	Prohibition from recreational use (e.g. swimming, fishing, kayaking), beach closure, impacts on aesthetics, cultural and spiritual values.
Agricultural productivity	Exclusion of contaminated water for irrigation results in increasing water scarcity. Irrigation with contaminated water causes damage to, and reduced productivity of, pasture and crops, contamination of soil, impacts to livestock health and production, and scouring of infrastructure.
Industrial productivity	Exclusion of contaminated water for industrial use results in increasing water scarcity. Scouring of infrastructure, and clean-up costs from spills/accidents.
Commercial fisheries	Direct and indirect fish kill, contamination of shellfish.
Urban and domestic use	Increased water treatment and inspection costs, maintenance costs from scouring and premature ageing of infrastructure, increased wastewater treatment costs with implementation of more strict regulations. Emergency and clean-up costs from spills/accidents.
Tourism	Losses in fishing, boating, rafting and swimming activities to other tourism activities or to other ventures with superior water quality.
Property values	Waterfront property values can decline because of unsightly pollution and odour.

Exposure to risks associated with poor water quality depends on a combination of variables related to:

- The characteristics of pollutants, individually and in combination, the characteristics of the receiving water body, timing, distance to source of pollution, and the stochastic environmental conditions (as illustrated in Figure 1.2, Chapter 1).
- The vulnerability to water quality risks, which depends on the extent of any historic pollution, access to treatment or alternative sources of usable water, and institutional and policy mechanisms, including response-time delays (both societal and ecological). For instance: different ecosystems will respond differently to pollution; pollution detection, social awareness, policy development and remediation actions will cause further delays depending on local resources (Figure 2.1); and the rate and extent of ecosystem recovery is not uniform (Falkenmark, 2011; Hipsey et al., 2015). For example, in parts of Canterbury, New Zealand, research has shown that 30 to 60 years' worth of nitrate in the soil has yet to reach the groundwater system, which will have further impact on Canterbury's drinking water supply and lowland stream quality (Webster-Brown, 2015).

- Multiple sources and pollutants, from multiple actors and sectors, operating in parallel, which complicates water risk assessments and policy responses to improve water quality (Falkenmark, 2011). For example, although each pollution source may have relatively little impact individually, their cumulative effect can be highly damaging.

Figure 2.1. **Response time delays in water pollution abatement (societal and ecological)**

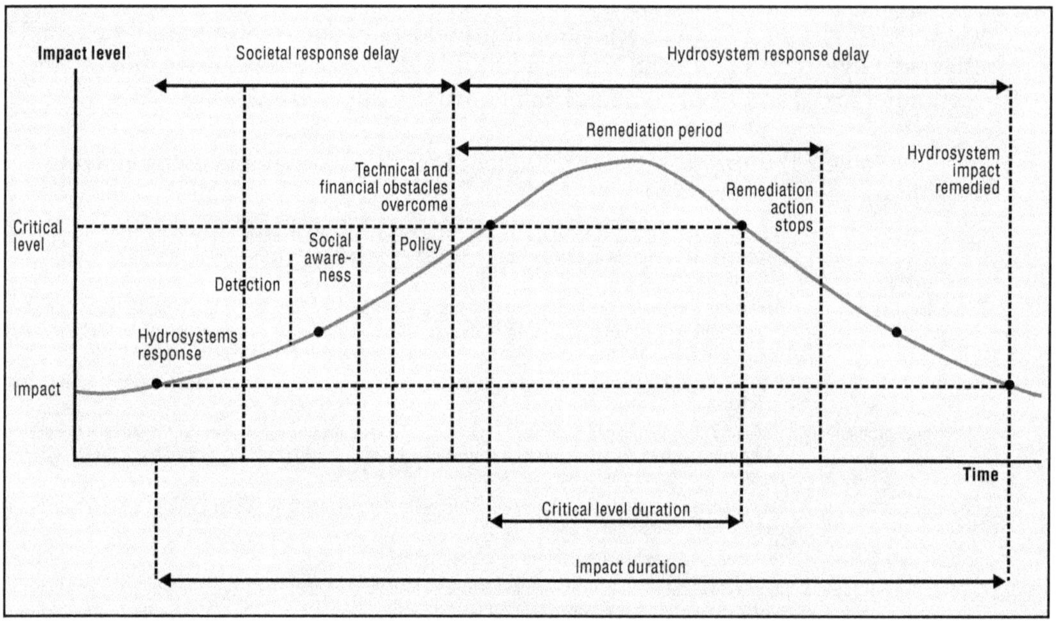

Note: The response times will be different for different ecosystems (hydrosystems), pollutants and local and national contexts.

Source: Falkenmark, M. (2011).

Market failure and water pollution

Economic theory suggests that under perfect conditions, markets will yield accurate incentives and foster efficient resource use. When particular conditions are not met, markets do not yield appropriate incentives and "fail" to achieve efficient resource use (Randall, 1983). Pollution of water resources is an example of market failure, an externality that is not accounted for in the market. For example, artificially low production costs in agriculture distorts the market and encourages over-production of food, feed and fibre that generates externalities such as nutrient runoff and eutrophication of water bodies that has economic, social and environmental costs to downstream users.

Moreover, water use is often a pure public good (non-rival and non-excludable, e.g. rivers) or a common pool resource (high rivalry, non-excludable, e.g. shared aquifers), which creates substantial transaction costs and can be subject to information gaps and uncertainties (Livingston, 1995). The market does not adequately supply public goods because private users cannot easily exclude non-paying beneficiaries and capture a return on investment. For example, it is not possible to exclude people living alongside a river from the benefits of improved water quality. The market also does not adequately provide for common pool resources because they are subject to overuse (high rivalry) in situations where strong common property resource institutions or resource user groups are not in place (OECD, 2015a).

Institutional (including regulatory, economic and voluntary policy instruments) and technical interventions are necessary to render efficient markets (Livingston, 1995) and internalise the negative externalities that lead to water pollution. The Polluter Pays Principle requires producers to pay the "full" cost (economic, social and environmental) of their production process, including externalities such as polluting water. The Beneficiary Pays Principle requires those who benefit from water quality improvements to pay for the costs incurred to do so.

However, in the event of market failure, public sector interventions or non-market approaches may not lead to the socially optimum solution. In many cases, non-market responses to market failures lead to less than optimal outcomes. This is in part because of the complex nature of water. For example, diffuse pollution is difficult to measure independently of the inputs that produced them, to pinpoint to individual land owners, and improved water quality can be difficult to prove/attribute to the uptake of best land management practices. Stock pollutants (with time delays in abatement measures spanning over more than one generation) and historic pollution (with those responsible no longer around) both pose complications in terms of who pays. Public policy distortions and spill-over impacts from other sector policies can also contribute to market failure. For example, policies aimed at protecting water quality may be at odds with other policies to increase and subsidise irrigation and intensive agricultural output for economic growth. Energy subsidies can encourage irrigation from groundwater sources, and cause saltwater intrusion with largely irreversible effects on groundwater quality.

It is therefore necessary for careful policy intervention to reduce the risk of market and government failure (such as policy distortions), and to overcome market imperfections such as uncertainty and information gaps that lead to negative impacts on water quality. A zero-pollution target is likely to be uneconomic and unaffordable (unless the risks are extreme), since the abatement cost of reaching it is likely to exceed the cost of the pollution itself. Instead, according to economic theory, the optimal pollution level from society's perspective is when the marginal abatement cost of pollution equals the marginal benefit from reducing the pollution level.

Economic valuation of ecosystems and water quality

Assessing the value, costs and benefits of water quality can assist policy makers in prioritising investments and determining policy options that provide the greatest potential societal benefit. For example, the debate over environmental protection is often about the trade-offs between the value of leaving areas in their natural state, and the opportunity costs of developing and exploiting them. Should a forest whose extensive root system reduces erosion, filters nutrients and provides many other environmental, social and recreational values be left un-cleared, or logged and converted to agriculture to contribute to economic growth, poverty alleviation and food security? Should wetlands that have high nutrient retention and the ability to remove bacteria, microbes, sediments and other pollutants, as well as providing other environmental, social and recreational values, be left in their natural state, or be drained and developed for housing, agriculture or other "productive" land uses? Economic valuation and cost benefit analysis (including determination of opportunity costs) can assist in answering such questions.

Economic valuation measures market and non-market values that people hold for freshwater ecosystems or for a certain standard of water quality. The concept of total economic value is a well-established and useful framework for identifying the various values associated with the environment (Table 2.2) (Atkinson and Mourato, 2015; OECD, 2006; IUCN, 1998).

Table 2.2. **Use values and non-use values of water ecosystems in relation to its quality**

Total economic value				
Use values			Non-use values	
Direct use values[1]	Indirect use values[2]	Option values[3]	Bequest values[4]	Existence values[5]
Drinking water Domestic use Agriculture Aquaculture, fisheries Energy production Industrial use Recreation Tourism Research Education	Nutrient retention Water purification / pollution abatement Habitat provision Climate regulation Soil erosion control	Future information Future uses (indirect and direct)	Values for legacy (future generations)	Freshwater biodiversity Ritual or spiritual values Culture, heritage Community values Aesthetic values Education and inspiration

Notes: 1. Value derived from direct human use of water of suitable water quality. 2. Value derived from the ecosystem services provided by freshwater ecosystems. 3. Value derived from the importance that people give to the future availability of freshwater of suitable quality for personal benefit (known or unknown), and for future value of information (e.g. untested genes of aquatic flora and fauna may provide future inputs into agricultural, pharmaceutical or cosmetic products). 4. Value attached by individuals to the fact that future generations will also have access to the benefits from species and ecosystems (intergenerational equity concerns). 5. Value related to the satisfaction that individuals derive from the mere knowledge that freshwater ecosystems and adequate water quality continue to exist.
Source: Adapted from IUCN (1998).

Determining the total economic value of the quality of a freshwater resource (use values and non-use values), whether it be a river, lake, wetland or groundwater aquifer, can be difficult, costly and time-consuming for a number of reasons:

- Although some of the negative externalities of degraded water quality are tangible, many are not, and their monetary quantification entails non-market valuation techniques (OECD, 2008). Even if there is a market value, this value may not reflect the "real" economic value due to market failure. For example, operating costs associated with wastewater treatment plants may not reflect the full social costs associated with pollution.
- Differentiating between point and diffuse sources of pollution through complex hydrological systems can be difficult. The separation of cause-and-effect by both physical distance and by time-lags adds complexity to the measurement and comparison of monetary values (OECD, 2012b).
- The "business as usual" scenario can be difficult to predict as many factors influence water quality. For example, the conversion of non-irrigated grassland to intensive irrigated dairy farming may occur with an increase in milk prices, and subsequently have a negative effect on diffuse nutrient pollution. Or the closure of a factory due to an economic downturn may lead to an improvement in water quality independent of policy intervention, as could the requirements of other legislation.
- Economics cannot fully account for all values (use and non-use values) attributed to a water resource. There are likely to be conflicting values, missing values and double counting identified during a total economic value study (IUCN, 1998). There are also uncertainties about the underlying environmental responses.

Furthermore, there is an array of different valuation methods which make comparison of different studies impossible. Such methods include: market prices, contingent valuation (willingness to pay or accept), hedonic pricing, travel cost method, change in productivity, loss (or gain) of earnings, opportunity cost, damage avoidance cost, and replacement cost, each of which have limitations. For example, contingent valuation is particularly controversial. There can be discrepancies between willingness to pay and willingness to accept: willingness to pay is often over-estimated due to loss aversion and the lack of requirement to actually pay; results are subject to survey design, instrument and starting point bias; and there may be limitations on information, and public education, risk perception and awareness that effect results (Hanley and Shogren, 2005).

The cost of water pollution and management options

Acknowledging the caveats around valuation in the previous section, the estimated cost of water pollution in OECD countries is substantial and attempts at estimating national costs have been reported in literature (Table 2.3). Note that a variety of methodological approaches have been used in the various studies, including in relation to whether they are reporting marginal, average or total costs. As such, results are difficult to interpret out-of-context, and cross-country or cross-study comparisons can be misleading. Furthermore, the true costs of pollution are also likely to be greater than the estimates suggest given the difficulty of calculating non-market values. Despite this, the valuation estimates serve to illustrate the existence of externalities and a need to adjust water management practices in order to reduce negative impacts on water quality.

Over many years, policies to address water pollution from agriculture across OECD countries, and to reduce the economic, environmental and social costs of pollution, have cost taxpayers in the order of billions of dollars annually and provided mixed results (OECD, 2012a). The lack of quantitative information about the benefits of reducing pollution is an obstacle to the formulation of efficient water quality objectives. Indeed, monetary estimates of the advantages are needed if water quality improvement targets are to balance costs and benefits. In an OECD survey on *Reducing water pollution and improving natural resource management,* six countries reported that a lack of information about benefits is a serious problem for water quality reform and recommended further efforts to better quantify them (Denmark, France, Italy, Sweden, Switzerland, United States) (OECD, 2005).

The fundamental challenge for policy makers is to develop water quality policy measures that can achieve environmental goals with the least overall costs for a given level of acceptable risk, including polluters' compliance costs and policy-related transaction costs, taking into account equity and other social factors (OECD, 2015b). Several studies compiled by the United States EPA (2015) have documented in-lake mitigation measures and their costs to remove nutrients from bottom sediments and the water column and their resultant algal blooms. For an individual water body, these costs range from USD 11 000 for a single year of barley straw treatment to more than USD 28 million in capital and USD 1.4 million in annual operations and maintenance for a long-term dredging and alum treatment plan (US EPA, 2015).

Table 2.3. Estimated annual national costs of water pollution: A selection from OECD countries

Country	Type of water quality impact	Annual cost (millions)			Source
		National currency	EUR	USD	
Australia	Algal blooms associated with excessive nutrients in freshwater	AUD 180 - 240	109 - 145	116 - 155	Atech, 2000
Belgium	Drinking water treatment costs		120 - 190	167 - 264	Dogot et al., 2010
France	Eutrophication of coastal waters (loss of tourism revenue and cost of cleaning up algae)		100 - 150	139 - 208	Bommelaer and Devaux, 2011
	Agricultural nitrate emissions and pesticides		610 - 1070	695 - 1219	Marcus and Simon, 2015
Korea	Reducing chemical contamination of drinking water			106	Kwak and Russell, 1994
Netherlands	Nitrate and phosphate pollution		403 - 754	371 - 695	Howarth et al., 2001
Spain	Nitrate and phosphate pollution		150	208	Hernandez-Sancho et al., 2010
Sweden	Coastal eutrophication Baltic Sea eutrophication		860 492 - 1466	1257 719 - 2143	Huhtala et al., 2009
Switzerland	Agricultural pollution	CHF 1000	608	690	Pillet et al., 2000
United Kingdom	Drinking water treatment costs, agricultural pollution of surface water, estuaries	GBP 229	335	458	Jacobs et al., 2008
England	Total cumulative cost of water pollution (point and diffuse sources)	GBP 700 - 1300	840 - 1560		National Audit Office, 2010
Europe	Human health and ecosystem impacts from nitrogen pollution of rivers and seas		40-155		Van Grinsven et al., 2013
	Health costs of nitrate in drinking water – colon cancer		1000		van Grinsven et al., 2010
United States	Freshwater eutrophication Protecting aquatic species from nutrient pollution Lakefront property values from nutrient pollution Recreational use from nutrient pollution		1500	2200 44 300 - 2800 370 - 1160	Dodds et al., 2009
	Drinking water impacts from nitrogen pollution Impacts of nitrogen pollution on freshwater ecosystems			19000 78000	Sobota et al., 2015
	Drinking water costs of nitrate contaminated wells			12000	Compton et al., 2011
	Pesticide contamination of groundwater		1610	2000	Pimentel et al., 2005
	Marine algal blooms		32 - 46	34 - 49	Anderson et al., 2000
	Cleaning up leaking underground petroleum storage tanks			800 - 2100	Nixon and Saphores, 2007
	Controlling highway runoff from major highways			2900 -15600	Nixon and Saphores, 2007
	Freshwater pollution by phosphorus and nitrogen			4300+	Kansas State University, 2008
	Health benefits of improving drinking water quality			130-2000	US EPA, 2006
	Costs of gastrointestinal illnesses attributed to drinking water			2100-1380	Garfield et al., 2003
	Health benefits associated with reducing arsenic from 50µg/L to 10 µg/L			140-198	US EPA, 2001
	Health benefits associated with reduction of nitrate exposure to legal safety standards			350	Crutchfield et al., 1997

Source: Updated from OECD (2012b); OECD (2008).

There are also substantial costs associated with restoring impaired waterbodies, such as developing: *i)* total maximum daily loads, minimum water quality standards or nutrient caps; *ii)* catchment management plans, and *iii)* pollution charges, or nutrient trading and pollution offset schemes (US EPA, 2015). For example, there are several trading and offset programmes in the United States that have been developed specifically to assist in nutrient reductions. One developed for the Great Miami River Watershed in Ohio for nitrogen and phosphorus had estimated costs of more than USD 2.4 million across 3 years (US EPA, 2015).

Prevention of diffuse pollution is often more cost effective than treatment/restoration options. In New Zealand, there are 18 recognised management practices that can reduce phosphorus losses from a range of farming enterprises, and each of them have varying cost-effectiveness (Table 2.4). Figure 2.2 shows the change in farm profit from the implementation of three mitigation measures to meet regional policy requirements in a catchment draining a dairy farm in southern New Zealand. In particular, soil testing to ensure optimal fertiliser applications is cost effective in reducing a) fertiliser costs, and b) nutrient losses associated with excess use.

Table 2.4. **Range of cost and effectiveness of management practices to mitigate phosphorus losses from New Zealand dairy farms**

Strategy	Main targeted P form(s)	Cost-range (USD/kg P conserved)	Effectiveness (% total P decrease)
Management			
Optimum soil test P	Dissolved and particulate	(highly cost-effective)[a]	5-20
Low P farming system	Dissolved and particulate	(330)	25-30
Low solubility P fertiliser	Dissolved and particulate	0-20	0-20
Steam fencing	Dissolved and particulate	2-45	10-30
Restricted grazing of cropland	Particulate	30-200	30-50
Greater effluent pond storage / application area	Dissolved and particulate	2-30	10-30
Flood irrigation management	Dissolved and particulate	2-200	40-60
Low rate effluent application to land	Dissolved and particulate	5-35	10-30
Amendment			
Tile drain amendments	Dissolved and particulate	20-75	50
Red mud (bauxite residue)	Dissolved	75-150	20-98
Alum to pasture	Dissolved	110->400	5-30
Alum to grazed cropland	Dissolved	120-220	30
Edge of field			
Grass buffer strips	Dissolved	20->200	0-20
Sorbents in and near streams	Particulate	275	20
Sediment traps	Dissolved and particulate	>400	10-20
Dams and water recycling	Particulate	(200)-400	50-95
Constructed wetlands	Particulate	100->400	-426-77
Natural seepage wetlands	Dissolved and particulate	100->400	<10

Notes: Numbers in parentheses represent net benefit, not cost. a) Depends on existing soil test phosphorus (P) concentration.
Source: McDowell et al. (2016).

Figure 2.2. **Change in median dissolved reactive phosphorus concentration and profitability 5 years after the consecutive implementation of three mitigation strategies in a catchment draining a dairy farm, Otago, New Zealand**

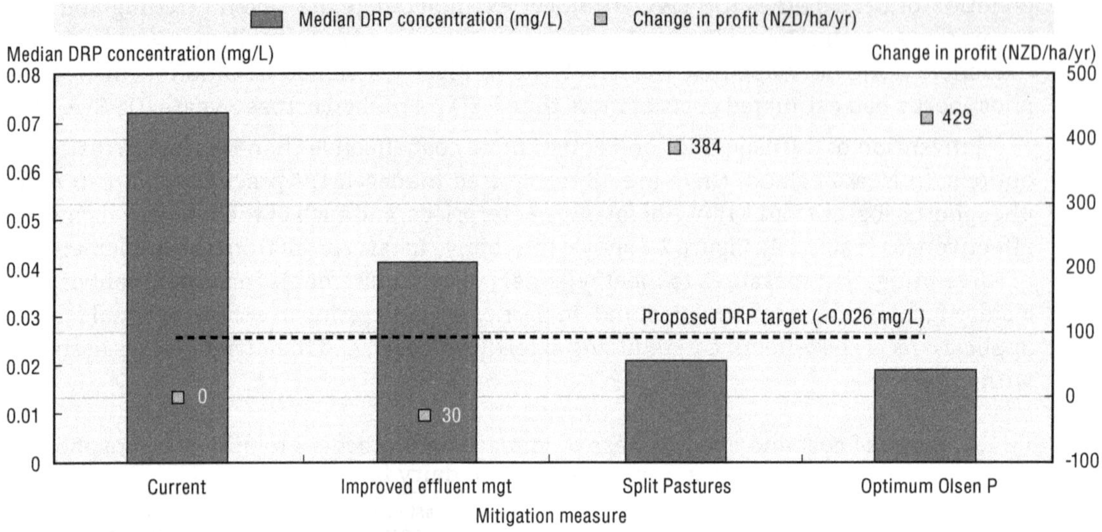

Note: Mitigation strategies to mitigate P loss a mixed model : 1) improving effluent management to prevent ponding and discharge of effluent-P through artificial drainage; 2) splitting mixed pastures into monocultures to have low-P requiring species in runoff producing areas with lower soil Olsen P; 3) lowering fertiliser inputs such that soil Olsen P test concentrations across the rest of the catchment that are no greater than the agronomic optimum. The proposed target refers to the Otago Regional Council policy requirement: <0.026 mg dissolved reactive P (DRP) per litre median concentrations in emissions at baseflow (thereby avoiding variability associated with stormflow).

Source: McDowell et al. (2016).

In a study of benefit to cost ratios of diffuse pollution mitigation measures to reduce agricultural pollutants (nitrogen, BOD, *E. coli* and *Cryptosporidium)* in the United Kingdom, Rothamsted Research (2005) found the most cost effective measures were characterised as relatively low cost and common sense solutions that work within the bounds of current agricultural practice. For example, the integration of manures with fertilisers when planning nutrient applications and avoiding spreading fertiliser at times of high risk. The measures that are least attractive are typically expensive when expressed as a cost-benefit ratio. For example, the reduction of livestock numbers on a farm or retiring land from production.

Modelling can help in determining the most cost-efficient mitigation measures and when and where to best apply them. Modelling demonstrates that the cost-effectiveness of mitigation practices tends to decrease the farther away from the pollution source a practice is implemented (e.g. McDowell and Nash, 2012). Targeting mitigation measures to the right place (i.e. vulnerable areas such as permeable soils, or land in close proximity to a water body) and at the right time (when losses from vulnerable areas are greatest, such as during high rainfall events and the rainy season) can increase farm profitability and reduce pollution mitigation costs. Mitigation measures targeted at one pollutant must also be assessed for potential impacts on other pollutants. For example, soil cultivation or tillage may reduce losses of dissolved reactive phosphorus, but may increase nitrogen losses.

The expected benefits (environmental, social and economic) of a policy response to address water pollution needs to outweigh its expected costs (environmental, social and economic). The cost of not intervening (Table 2.3) must also be taken into consideration in the analysis. Whether policy responses bring larger benefits than costs is an empirical question and has to be examined for each case. Natural capital accounting has the potential to be an effective tool in assessing the costs and benefits of protection of freshwater ecosystems and improvements in water quality. Experience from the United Kingdom is described in Box 2.1. Advances in this area will improve water quality valuation in the future.

Box 2.1. Natural capital accounting as a tool to value natural resources and ecosystem services: Experience from the United Kingdom

Natural capital accounting (NCA) provides a basis for valuing natural capital assets, and the ecosystem services they provide, by quantifying the "costs and benefits" of resource management decisions (Clothier et al., 2013; Mackay et al., 2011). Profit and loss statements reflect the cost of externalities of consuming natural resources, and investment in natural capital can be evaluated and weighed up against investment in engineering solutions. Ecological economics and NCA can also guide issues of sustainable development, intergenerational equity, irreversibility of environmental change, uncertainty of long-term outcomes and exploitation of natural resources for short-term profit (Faber, 2008).

The United Kingdom is experimenting with using NCA. They face issues of water scarcity in places, high demand and environmentally unsustainable abstraction in certain regions; and surface and groundwater quality problems from diffuse agricultural and urban pollution, in particular high levels of nitrates and pesticides, despite major investment in point source control. The NCA approach naturally aligns with a catchment-scale approach and demonstrates that multiple benefits can be derived from investing in ecosystems services and natural capital, such as forests, floodplains and wetlands.

The Office for National Statistics (ONS, 2016) uses natural capital accounting to:

- Quantify the losses, gains and relative importance of services provided by natural assets; the development of monetary accounts enables the value of different services to be monitored and comparisons to be made with the value of other economic assets.
- Highlight links with economic activity and pressures on natural capital.
- Inform priorities for resourcing and management decisions.

In a first attempt to develop initial experimental statistics on UK freshwater ecosystem assets and ecosystem services, estimates of the monetary values of UK wetlands and open channels were based on a number of indicators, and the condition of freshwaters between 2008 and 2012 (Khan and Din, 2015). The monetary value of UK freshwaters was estimated at a total of GBP 39.5 billion in 2012, 10% higher than in 2008 (this was mainly due to an increase in the monetary value of UK open waters) (Khan and Din, 2015). These estimates exclude other valuable services such as the traded price of electricity generated by hydropower, which was over GBP 300 million in 2012; GBP 8 million worth of navigation licences, which were issued in England and Wales in 2012/13; and landscape amenity values, which are also important benefits (e.g. property price premiums in close proximity to canals and rivers).

Table 2.5 **Asset values of UK freshwater ecosystems (wetlands and open channels) 2008-12**

(GBP, 2012 prices)

	Freshwater ecosystems GBP billion				
Services	**2008**	**2009**	**2010**	**2011**	**2012**
		Provisioning services			
Fish extraction	1.1	0.9	0.8	1.2	0.9
Water abstraction	18.5	18.7	19.3	20.8	23.9
Peat extraction	0.3	0.3	0.3	0.3	0.2
		Cultural services			
Recreational visits	16.2	15.8	14.9	14.9	14.5
Educational visits[1]	0.0	0.0	0.0	0.0	0.0
Total	36.1	35.8	35.3	37.1	39.5

Note: Results are a gross underestimate of the true value of UK freshwater assets and ecosystems services due to limitations with data. A number of freshwater ecosystems are not included in this valuation. 1. The actual figures for the NPV (in GBP) of educational visits are as follows; (2008) 1.0 million, (2009) 1.0 million, (2010) 1.0 million, (2011) 1.0 million and (2012) 0.9 million.

Sources: Clothier et al. (2013); Faber (2008); Khan and Din (2015); Mackay et al. (2011); ONS (2016); Water UK (2013).

Table 2.6. **Regulatory, economic, and voluntary pollution control mechanisms, directed at source and end-of-pipe**

	Regulatory approaches	Economic instruments	Voluntary / Information instruments
Source-directed approaches	**Standards** E.g. Planning requirements (i.e. environmental impact assessments, nutrient accounting, nutrient management plans, protection zones) Mandatory use of best management practices (i.e. manure storage, riparian zones) Restrictions or bans on the use of chemicals Pesticides and hazardous chemicals registration Cap on modelled diffuse pollution Restrictions on pesticides, fertiliser, manure, effluent, biosolids application rates and timing Input quotas per hectare and restrictions on livestock densities Land retirement requirements (such as riparian buffer strips) Standards to induce the use of new tools, including monitoring and information communication technology. **Liability rules** E.g. Negligence liability rules **Performance labelling** Government requirements for labelling on level of environmental performance	**Taxes** E.g. Taxes on chemical and solvent purchases Taxes on fertiliser or pesticide purchases Taxes on manure applications Taxes on domestic products such as personal care products, detergents **Subsidies** Subsidies that reward inputs, practices, or technologies that prevent pollution Agricultural land retirement subsidies Subsidies for R&D or to induce uptake of new technologies, including monitoring technologies. **Payment for Ecosystem Services** E.g. payment by downstream users to upstream users, in exchange for practices that reduce pollution and protect water quality.	**Contracts/Bonds** E.g. Land retirement contracts Contracts involving the adoption of conservation practices Contracts involving the adoption of nutrient management practices **Best Environmental Practices** E.g. best practices for fertiliser and pesticide applications to reduce runoff **Advisory services** E.g. Farm advisory services and demonstration projects to encourage greater uptake of best environmental practices and improve productivity **Environmental labelling** Products that meet certain environmental standards can be marketed and sold at a premium and/or subsidised. **Corporate Social Responsibility** Investment in practices that reduce pollution to improve corporate image, water stewardship metrics and ISO water footprint standards. Covenants and negotiated agreements Industry code of conduct Private standards (e.g. food and beverage companies requiring suppliers to comply with certain environmental conditions) **Benchmarking** Publicising and ranking polluter's performance **Self-regulation** Polluters acting to regulate themselves Community-based regulation and co-operation agreements **Research and knowledge building** Private and public research to improve understanding of water quality risks Knowledge sharing and problem solving at the community level **Information campaigns** E.g. targeted at households to reduce disposal of chemical waste and unused pharmaceuticals in toilets
End-of-pipe approaches	**Standards** E.g. Permits for discharges with quantity and quality conditions Restrictions on modelled diffuse pollution (i.e. nutrient loadings) **Non-compliance penalties and fines** Non-renewal of resource permits or greater restriction on current permits	**Taxes** E.g. Taxes on volume of point source discharges Taxes on modelled diffuse pollution (i.e. nutrient loadings) Taxes on estimated soil loss **User charges** Sewer surcharge (can incentivise reductions in wastewater from businesses and households and raise revenue to finance wastewater treatment plant upgrades) **Markets** Water quality trading of point discharge permits Water quality trading of modelled diffuse pollution discharge permits Point-non-point trading **Loans** E.g. For investment in WWTPs or artificial wetlands **Subsidies** Subsidies for inputs, practices, or technologies that reduce pollution Subsidies for R&D or to induce uptake of new technologies, including monitoring technologies **Water quality offsets** Liability for pollution and payment for compensation of damage	**Cost-sharing programmes** E.g. PPPs for investment in wastewater treatment plants or decentralised wastewater reuse systems
Cross-sector approaches	**Minimum environmental flows** (for pollution dilution)	**Removal of harmful subsidies** **User charges** Water charges (marginal cost pricing can reduce excessive water use and consequent pollution; and can also raise revenue for drinking water treatment plant upgrades) **Taxes** Taxes on atmospheric pollutant emissions (which can lead to water pollution, i.e. acidification)	

Sources: Adapted from Shortle and Horan (2001); Metz and Ingold (2014).

Generally, reductions in pollution loads, typically by regulatory and voluntary approaches, in OECD countries appear not to have been done so at the lowest cost (Shortle et al., 2012; NAO, 2010; OECD, 2005). The costs of regulation and control of point source pollution have been sizeable in many OECD countries. For example, in the United States, the Environmental Protection Agency considers that the Clean Water Act, which regulates point source discharges through a permitting system, has provided benefits in line with costs. But other authors have found benefit-cost ratios of 1:6 (Freeman, 2003) or even 1:20 for those Clean Water Act regulations that have been subject to regulatory impact assessment between 1981 and 1996 (Hahn, 2000b). Olmstead (2010) argues that the Clean Water Act brought net benefits up to the late 1980s, but that afterwards the incremental costs exceeded the incremental benefits. The costs of preventing the degradation of water quality should receive similar attention.

Appropriate targeting of both point source and diffuse source pollution are key factors that must be taken into account to overcome policy inefficiencies. There is a case for the utilisation of cost-effective prevention and abatement practices that could yield more beneficial results in terms of water quality improvements and control-cost savings (Shortle et al., 2012; Shortle and Horan, 2013). Innovative approaches, such as water quality trading and other economic instruments, offer the possibility of improving the effectiveness and efficiency of water quality programmes (OECD, 2015b; Shortle and Uetake, 2015).

Water pollution control mechanisms in OECD countries

Government intervention to reduce pollution typically takes any one, and almost always a combination of, three basic forms:

1. *Regulation:* setting performance or technology standards to reduce pollution. For example, quantitative limits on the quality and volume of discharge may be specified by a permit and enforced by law. Technology standards may include the specification of minimum technological standards for wastewater treatment plants, such as a requirement for tertiary treatment and the removal of nutrients, using technology such as biological nutrient removal, sand filters or chemical precipitation. There may bans on certain harmful chemicals, restrictions on land use activities and mandatory use of best management practices.
2. *Economic instruments:* taxing environmentally harmful products, pollution charges on emissions, providing economic incentives (i.e. subsidies), and designing tradable permits, to reduce pollution and negative externalities, and/or raise revenue to pay for research, mitigation, adaptation and treatment of poor water quality.
3. *Voluntary or information instruments:* guiding and supporting households, farmers or industry to reduce pollution voluntarily. For example, farmers may be encouraged to fence off stock access to water courses and to provide a riparian strip aimed at interrupting the movement of diffuse nutrient and sediment transfer from agricultural land to surface waters. At the catchment and sub-catchment scale, resource users and industry can be supported in planning for better collaborative outcomes. Households may be informed and encouraged to reduce their impacts by using alternative cleaning products, such as low-phosphate laundry detergent.

With water quality continuing to deteriorate, OECD countries are progressively looking to manage diffuse pollution sources (e.g. the EU Water Framework Directive, US Clean Water Act and New Zealand's National Policy Statement for Freshwater Management). Examples of regulatory, economic, and voluntary pollution control methods are presented in Table 2.6. They are distinguished between source-directed approaches (targeted at reducing or preventing the source of pollution) and end-of-pipe approaches (targeted at reducing the impact of pollution). Chapter 3 will discuss in more detail selected policy instruments to manage diffuse water pollution.

Determining who pays for pollution abatement and water quality improvements

The three primary actors to cover the costs of providing water quality management are: *i)* the polluter; *ii)* the beneficiary; or *iii)* government as outlined below:

- *The Polluter Pays Principle* creates conditions to make pollution a costly activity and to either influence behaviour to reduce pollution, or generate revenues to alleviate pollution and compensate for social costs (OECD, 2012a). Examples include pollution charges, taxes on inputs (such as fertilisers and pesticides), and sewer user charges.
- *The Beneficiary Pays Principle* allows sharing of the financial burden of water quality management. It takes account of the high opportunity cost related to using public funds for the provision of private goods that users can afford. A requisite is that private benefits attached to water resources management are inventoried and valued, beneficiaries are identified, and mechanisms are set to harness them. (OECD, 2012a).
- *Public budgets* (i.e. from general taxation) often cover the costs of providing water quality management functions that serve the public more generally (OECD, 2015c).

There are several challenges that result in the Polluter Pays Principle not frequently being applied in the control of diffuse pollution (it is more commonly used with the control of point source pollution) (Table 2.7). However, despite strong political opposition from polluters, in the instances where high levels of taxes have been applied to inputs to comply with the Polluter Pays Principle, often coupled with a mix of other policy measures, they have usually led to reductions in input use without loss of farm production or income (OECD, 2012c).

The Beneficiary Pays Principle is used to some extent to control diffuse pollution from agriculture, usually on a voluntary basis. For instance, through Payment for Ecosystem Services schemes from downstream utilities to upstream farmers in exchange for land management practices that reduce pollution (OECD, 2015d). However, equity concerns can arise if payments are seen to "reward polluters" while neglecting producers already demonstrating best practice (OECD, 2013). Requiring that minimum regulatory standards to reduce pollution be met before payments are made is one way to overcome equity issues, combining both the Polluter Pays and Beneficiary Pays Principles. A summary of the advantages and challenges of the Polluter Pays and Beneficiary Pays Principles for the control of diffuse pollution are outlined in Table 2.7.

Table 2.7. **Advantages and challenges with the Polluter Pays and Beneficiary Pays Principles for the control of diffuse pollution**

	Polluter Pays Principle	Beneficiary Pays Principle
Advantages	Internalises the external cost of pollution. Provides an opportunity to prevent pollution. Provides an opportunity to adopt best management practices. Provides an opportunity to raise revenue for water quality management. Demonstrated success with high level input taxes have led to reduction in input use and water pollution without loss of farm production.	Allows sharing of the financial burden of water quality management with users that benefit. Provides an opportunity to adopt best management practices. Provides an opportunity to raise revenue for water quality management. Demonstrated success with payment for ecosystem services by downstream utilities to upstream farmers in return for land management practices that reduce pollution.
Challenges	Poor enforcement of existing regulations on diffuse pollution. Diffuse pollution sources are not easily directly measured at reasonable cost with current monitoring technologies (although computer modelling is a cost-effective alternative). Difficulty with identifying and targeting the polluters. Undefined property rights. High transaction costs associated with multiple polluters. Difficulty with determining reliable estimates of potential costs and benefits. Strong political opposition from polluters.	Seen as "rewarding" the polluter. Beneficiaries of water-related services do not usually pay the full cost of the provision of ecosystems, or may free ride. Difficulty with determining reliable estimates of potential costs and benefits. Private financiers are not guaranteed to benefit from payments and may have a reduced incentive to support them: changes in land use management may not lead to water quality benefits, long time-lag before improvements are visible, landowners or their managers may not comply. Difficulty with identifying and targeting the polluters. Undefined property rights.

Sources: OECD (2015c,d; 2013; 2012a,b); Smith and Porter (2010).

When developing policy to manage water quality, an important consideration is not only the measurement of the costs and benefits of water pollution reductions, but also on to whom these costs and benefits will fall. Box 2.2 outlines under which type of policy instrument the costs fall upon.

Box 2.2. **Who pays for, and who receives the benefits of, water quality improvements?**

Water quality improvements come at a cost, and those benefitting from improvements in water quality are not necessarily those who pay for the cost of pollution reduction, and those that pollute do not necessarily pay damage costs. For example, diffuse pollution from agriculture is loading costs onto other sectors as well as the environment. Who bears the costs and reaps the benefits of water quality improvements typically depends on the policy instrument used:

- *Regulations, taxes and markets*: improvements in water quality are usually at the cost of the polluter, the costs of which can be passed onto the consumer.
- *Economic subsidies and incentives*: improvements in water quality are at the cost of the tax payer.
- *Environmental labelling and Corporate Social Responsibility*: improvements in water quality are at the cost of producers, and corporations who sell and manage commercial goods. The cost is ultimately passed onto the consumer.
- *Payment for Ecosystem Services*: changes in management practices that improve water quality are at the direct cost of the beneficiaries.

Without effective policy instruments to reduce pollution, the cost of pollution typically falls on drinking water utilities (and subsequently households) and downstream water users, such as downstream industry and agricultural users, eco-tourism operators, recreational users, and waterfront property owners.

The need for policy coherence

There is a need for coherence across a number of policies - agriculture, energy, industry, economic, spatial and urban planning, waste, construction, transport, climate, and environment (water, air, land and biodiversity). For example, a management approach considering air, water and land management practices to manage nitrogen loading has the potential to be more cost-effective and provide environmental co-benefits[1], as illustrated in Box 2.3.

Box 2.3. **An example of the importance of policy coherence for water quality, Chesapeake Bay, United States**

Agricultural runoff is not the only source of diffuse nitrogen pollution to impact waterways, fallout of atmospheric pollutants also contribute: nitrogen oxides (NOx), sourced primarily from energy and transport emissions, and ammonia (NH_3) sourced primarily from agriculture (in particular, volatilisation of stock manure and effluent).

In one scientific study of Chesapeake Bay, Birch et al. (2011) showed that the damage costs from nitrogen air emissions are much larger in the watershed than those from the whole of land and water emissions. The case demonstrates that a unique focus on metrics relating to the Bay water quality could result in a missed opportunity to prioritise actions on air quality that could have larger benefits, including health, throughout the entire chemical cascade and a much broader geographical area.

In essence, the reduction of Chesapeake Bay damages from nitrogen loading (including freshwater and estuarine impacts) may benefit more from a more strict control of air pollution (at the airshed level) rather than from more strict water pollution controls (at the watershed level) (OECD, forthcoming). Thus, a management approach considering air, water and land management practices to manage nitrogen loading has the potential to be more cost-effective and provide environmental co-benefits.

Sources: Birch M.B. et al. (2011); OECD (forthcoming).

Policy coherence is also required to avoid conflicting signals and incentives, in particular to farmers in achieving sustainable water management (Parris, 2012). Some government non-environmental programmes and subsidies inadvertently work in opposition to efforts to improve water quality. For example, policies that support agriculture production encourage greater land use change and intensive use of inputs, such as fertilisers, pesticides, irrigation, and fossil fuel use (Shortle et al., 2012). Input subsidies can also encourage more intensive use of potentially environmentally harmful inputs. An example of perverse incentives causing an increase in water pollution is presented in Box 2.4.

As water quality is intrinsically linked with water quantity, and both are affected by climate change, the use of various policy instruments in urban water management, and the food, energy, and industrial sectors, can have wider environmental and social impacts. Cities can also be a part of the solution. For example, taxes on impervious surfaces in urban areas can incentivise reductions in stormwater runoff and finance a greater proportion of urban land to be connected to a drainage system with stormwater treatment. In Austin, Texas, drainage fees are used to reduce risks of flash flooding, erosion and water pollution (City of Austin, 2016). In Santa Monica, California, stormwater property taxes are used to fund the city's watershed management programme and it's obligation to comply with federal and state Clean Water Act regulations (City of Santa Monica, 2016).

Box 2.4. Perverse subsidies for bioethanol production, United States

While biofuels may yield renewable fuel benefits, there could be downsides in terms of water quality and other environmental stressors.

US government support for ethanol production has resulted in higher corn prices globally. In turn, this has incentivised intensification and extensification of corn production. As a result, nutrients and pesticides entering US water bodies has increased, and conservation land has decreased. Increased corn production in the Midwest of the US is also thought to have increased nutrient loading to the Gulf of Mexico.

Higher corn prices attributable to the US ethanol policy are estimated to reduce land offered into the US Conservation Reserve Programme by about 5%. In addition, about a third of land currently enrolled in the programme is likely opt-out to take advantage of higher corn prices if there were no penalties for doing so. This not only impacts biodiversity, but also reduces the water filtration ecosystem services that conservation land provides. Furthermore, despite the good intentions of biofuels to reduce greenhouse gas emissions and mitigate climate change, global emissions of greenhouse gases are likely to have actually increased as ethanol policies may have created global changes in land use by reducing forest land and grassland that act as greenhouse gas sinks.

Sources: Hanley et al. (2013); Hellerstein and Malcolm (2011); Searchinger et al. (2008); Secchi et al. (2011); Shortle et al. (2012).

Special consideration will need to be given to policy design to ensure that policies are responsive and flexible in adapting to future changes, and particular consideration be given to the effects of climate policies on water quality. The following section elaborates on the potential trade-offs and the need for policy coherence between climate and water quality policies. Similarly, there will also be trade-offs between policies with other sectors, in particular agriculture and energy.

The potential effects of climate change mitigation and adaptation policies on water quality

Some measures to reduce greenhouse gases and adapt to climate change may conflict with existing efforts and regulations to improve and maintain water quality (Fezzi et al., 2015; IPCC, 2014; OECD, 2014). Such trade-offs have received little consideration in decision-making, policy design and academia (Fezzi et al., 2015). For example, climate adaptation by way of increased irrigated agriculture can generate significant adverse impacts on water quality. In Great Britain, the area of land at risk of high nutrient loading in water bodies is projected to increase by 30-40% as a result of climate change adaptation measures taken by farmers (Fezzi et al., 2015). Hydropower can modify the natural flow of watercourses, which may have some effects on capacity for dilution of point source pollution and on freshwater ecosystems. Carbon capture and storage may decrease groundwater quality, causing acidification of freshwater aquifers due to leakage of pipes, and salt water intrusion of freshwater aquifers due to displacement of saline formations (IPCC, 2014).

A summary of the potential trade-offs and co-benefits between climate change mitigation/adaptation measures and water quality are presented in Table 2.8. The impacts on water quality demonstrate the importance of anticipating trade-offs and the wider impacts of climate change mitigation and adaptation measures when designing environmental policies. Conversely, when climate change mitigation and adaption policies are well-designed and the trade-offs managed, they can produce multiple benefits for food and water security, human health, air and water quality, and natural resource management. For example, soil carbon sequestration can increase nutrient and water retention and resilience to droughts and flooding (OECD, 2014). Riparian vegetation can provide cooling effects on in-stream temperatures (Davies, 2010), as well as enhance biodiversity, provide erosion control and filter pollutants (Pittock, 2009).

In order to create the conditions needed for win-win outcomes, better integration and implementation of policies and programs at all scales is required (Measham et al., 2011; IPCC, 2014). Furthermore, more studies are required to establish baselines to isolate the water quality impacts derived from climate change from other anthropogenic pollution causes, and to assess the vulnerability and ways of adapting to those impacts (IPCC, 2014). Modelling of various scenarios can offer a way forward but management decisions need to account for uncertainties around climate change projections regionally and locally, and the impacts on water quality.

Table 2.8. **Examples of potential trade-offs and co-benefits between climate change mitigation/adaptation measures and water quality**

Climate change mitigation	Climate change adaptation
Trade-offs	***Trade-offs***
Construction of hydropower dams and dykes can impact freshwater ecosystems, contribute to thermal pollution and may cause some sediment loads when sediment is released.	Increased intensive and irrigated agriculture can lead to greater loads of nutrients, pathogens, pesticides, sediment and other pollutants, and reduce environmental flows in rivers for pollution dilution.
Bioenergy crops increase water demand and decrease water quality and food security.	Increased groundwater abstraction can cause salt water intrusion, and if used for irrigation, can cause salinization of soils and groundwater contamination from leaching of nutrients, pathogens, pesticides and other pollutants.
Carbon capture and storage can decrease groundwater quality, causing acidification of freshwater aquifers due to leakage of pipes, and salt water intrusion of freshwater aquifers due to displacement of saline formations.	Construction of irrigation dams and dykes can impact freshwater ecosystems and cause a sudden flux of sediment when sediment released. Impounded storage may have some effects on environmental flows and capacity for dilution of point sources.
Concentrated solar power as a form of renewable energy requires substantial volumes of water for cooling (EWEA, 2014) and therefore can contribute to thermal pollution of water bodies.	Urban wetlands designed primarily for flood control can promote mosquito breeding.
Fracking for shale gas can contaminate aquifers with hydrocarbons and other toxic pollutants, and can lead to groundwater salinisation, potentially putting drinking water supplies at risk.	
Co-benefits	***Co-benefits***
Conservation and afforestation of water catchments reduces soil erosion and local flood risk, and improves water quality (nitrogen, phosphorus, suspended sediments) and instream habitat quality.	Precision agriculture (irrigation and nutrients) and water use efficiency can improve water quality.
Conservation of wetlands provides multiple benefits, such as water purification, water supply, flood regulation and biodiversity.	Soil carbon sequestration and building up organic matter can improve soil moisture and nutrient retention.
Stacking water quality credits with carbon credits increases the profitability of carbon sequestration practices, and vice versa. For example, the provision of CO_2-eq offsets through reductions of nitrogen fertiliser applications or through the establishment of green riparian buffer strips may not profitable without water quality credits. Allowing stacking increases participation and uptake of mutually beneficial environmental practices.	Development of adaptation plans to maintain optimal farming conditions are also an opportunity to combine with water quality objectives.
Soil carbon sequestration and build-up of soil organic matter can increase the capability of the soil to filter and retain water and nutrients. Conservation tillage and agroforestry can reduce soil erosion and improve water quality.	Wastewater reuse can provide a reliable source of water for irrigation, and returns nutrients to the land without the need to discharge effluent directly to water bodies or use synthetic fertilisers.
Precision agriculture (irrigation and nutrient) can reduce runoff of nutrients. Nitrogen inhibitors can reduce nitrate losses to water bodies.	Water efficient cultivars reduce water consumption and therefore potentially water pollution.

Sources: EWEA (2014); Lankoski et al. (2015); Millennium Ecosystem Assessment (2005); OECD (2014).

Note

1. Co-benefits from the water quality perspective: additional benefits beyond improvements in water quality resulting from pollution mitigation measures. Examples of potential co-benefits may include improved ecosystem and human health, reduced greenhouse gas emissions, improved air quality etc.

References

Atkinson, G. and S. Mourato (2015), "Cost-Benefit Analysis and the Environment", OECD Environment Working Papers, No. 97, OECD Publishing, Paris, http://dx.doi.org/10.1787/5jrp6w76tstg-en.

Birch, M.B. et al. (2011), "Why Metrics Matter: Evaluating Policy Choices for Reactive Nitrogen in the Chesapeake Bay Watershed", *Environmental Science & Technology.*, Vol. 45 (1).

Bommelaer, O. and J. Devaux (2011), Assessing water pollution costs of farming in France, Department of the Commissioner General for Sustainable Development, No. 52 EV English version.

City of Austin (2016), Drainage Charge: What is it and why is it important? (website), Watershed Protection Department, City of Austin, www.austintexas.gov/faq/drainage-charge-what-it-and-why-it-important (accessed 06/11/2016).

City of Santa Monica (2016), Urban Runoff: Stormwater Parcel Fees (website), Office of Sustainability and the Environment, City of Santa Monica, www.smgov.net/Departments/OSE/Categories/Urban_Runoff/Stormwater_Parcel_Fees.aspx (accessed 06/12/2016).

Clothier, B. et al. (2013) 'Natural Capital - thinking about how we value and use our natural resources', *AgScience*, 43(March 2013).

Crutchfield, S. R. et al. (1997), "Benefits of Safer Drinking Water: The Value of Nitrate Reduction", *American Economic Review*, June 1997.

EWEA (2014), Saving water with wind energy, European Wind Energy Association, Bruxelles, Belgium.

Faber, M. (2008) 'How to be an ecological economist', *Ecological Economics*, 66(1), 1-7.

Falkenmark, M. (2011), Water-a reflection of land use: Understanding of water pathways and quality genesis, International Journal of Water Resources Development, Vol. 27 (1), pp. 13-32.

Fezzi, C. et al. (2015), The environmental impact of climate change adaptation on land use and water quality, *Nature Climate Change*, Vol. 5, pp. 255–260.

Freeman, A.M. (2002), Environmental policy since Earth Day I: what have we gained?, *Journal of Economic Perspectives*, Vol. 16, No. 1, pp. 125–146.

Garfield, S.J. et al. (2003), "Public Health Effects of Inadequately Managed Stormwater Runoff", *American Journal of Public Health*, Vol. 93, No. 9, pp. 1527-1533.

Hahn, R.W. (2000), "Reviving regulatory reform: a global perspective", *AEI-Brookings Joint Centre for Regulatory Studies Working Paper*, Washington DC.

Hanley, N., Shogren, J.F. and E. White (2013), The Economics of Water Pollution, In: *Introduction to Environmental Economics*, second Edition, Oxford University Press, Oxford, UK.

Hanley, N. and J.F. Shogren (2005), Is Cost–Benefit Analysis Anomaly-Proof?, *Environmental & Resource Economics*, Vol. 32, pp. 13–34.

Hellerstein, D. and S. Malcolm (2011), *The Influence of Rising Commodity Prices on the Conservation Reserve Program*, ERR 110, United States Department of Agriculture, Economic Research Services.

Hipsey, M.R. et al. (2015), Predicting the resilience and recovery of aquatic systems: A framework for model evolution within environmental observatories, *Water Resources Research*, Vol. 51 (9), pp. 7023-7043.

IPCC (2014), *Climate Change 2014: Impacts, Adaptation, and Vulnerability. Part A: Global and Sectoral Aspects.* Contribution of Working Group II to the Fifth Assessment Report of the Intergovernmental Panel on Climate Change [Field, C.B. et al. (eds.)]. Cambridge University Press, Cambridge, United Kingdom and New York, NY, USA, 1132 pp.

IUCN (1998), *Economic Values of Protected Areas: Guidelines for Protected Area Managers*, prepared by the Task Force on Economic Benefits of Protected Areas of the World Commission on Protected Areas (WCPA) of IUCN, in collaboration with the Economics Service Unit of IUCN, IUCN, Gland, Switzerland and Cambridge, UK. 62 pp.

Khan, J. and F. Din (2015), *UK Natural Capital – Freshwater Ecosystem Assets and Services Accounts*, Office for National Statistics, http://webarchive.nationalarchives.gov.uk/20160105160709/www.ons.gov.uk/ons/dcp171766_398822.pdf [accessed 04/06/2016].

Kwak, S. J. and C. S. Russell (1994), "Contingent Valuation in Korean Environmental Planning: A Pilot Application to the Protection of Drinking Water Quality in Seoul", *Environmental and Resources Economics*, Vol. 4, pp. 511-526.

Lankoski, J. et al. (2015), "Environmental Co-benefits and Stacking in Environmental Markets", *OECD Food, Agriculture and Fisheries Papers*, No. 72, OECD Publishing, Paris. http://dx.doi.org/10.1787/5js6g5khdvhj-en.

Livingston, M. L. (1995), Designing Water Institutions: Market Failures and Institutional Response, *Water Resources Management*, Vol. 9, pp. 203-220.

Mackay, A. D. et al. (2011) 'Land: Competition for future use', *New Zealand Science Review*, 68(2).

Marcus, V. and O. Simon (2015), "Les pollutions par les engrais azotés et les produits phytosanitaires: coûts et solutions» (Pollution from nitrogen fertilisers and plant protection products: costs and solutions), du Service de l'Économie, de l'Évaluation et de l'Intégration du Développement Durable (SEEIDD) du Commissariat Général au Développement Durable (CGDD), No. 136, France.

McDowell, R.W. et al. (2016), A review of the policies and implementation of practices to decrease water quality impairment by phosphorus in New Zealand, the UK, and the US, *Nutrient Cycling in Agroecosystems*, Vol. 104, pp. 289–305.

McDowell, R.W. and D. Nash (2012), A review of the cost-effectiveness and suitability of mitigation strategies to prevent phosphorus loss from dairy farms in New Zealand and Australia, *Journal of Environmental Quality*, Vol. 41, pp. 680–693.

Measham, T.G. et al. (2011), Adapting to climate change through local municipal planning: barriers and challenges, *Mitigation and Adaptation Strategies for Global Change*, Vol. 16, pp. 889–909.

Metz, F. and Ingold, K. (2014), Sustainable Water Management: Is it possible to regulate micropollution in the future by learning from the past? A policy analysis, Sustainability, Vol. 6, pp.1992-2012.

Millennium Ecosystem Assessment (2005), *Ecosystems and Human Well-being: Biodiversity Synthesis. World Resources Institute*, Washington, DC.

National Audit Office (2010), *Tackling diffuse water pollution in England*, Report by the Comptroller and Auditor General, UK, July 2010.

Nixon, H. and J-D. Saphores (2007), Impacts of Motor Vehicle Operation on Water Quality in the United States: Clean-up Costs and Policies, University of California Transportation Center.

OECD (forthcoming), Forthcoming report on The Human Impacts on the Nitrogen Cycle, OECD Studies on Water, OECD Publishing, Paris.

OECD (2015a), *Water Resources Allocation: Sharing Risks and Opportunities*, OECD Studies on Water, OECD Publishing, Paris. http://dx.doi.org/10.1787/9789264229631-en.

OECD (2015b), *Public Goods and Externalities: Agri-environmental Policy Measures in Selected OECD Countries*, OECD Publishing, Paris, http://dx.doi.org/10.1787/9789264239821-en.

OECD (2015c), *Fostering Green Growth in Agriculture: The Role of Training, Advisory Services and Extension Initiatives, OECD Green Growth Studies*, OECD Publishing, Paris, http://dx.doi.org/10.1787/9789264232198-en.

OECD (2015d), *Water and Cities: Ensuring Sustainable Futures, OECD Studies on Water*, OECD Publishing, Paris. http://dx.doi.org/10.1787/9789264230149-en.

OECD (2014), *Climate Change, Water and Agriculture: Towards Resilient Systems*, OECD Studies on Water, OECD Publishing, Paris, http://dx.doi.org/10.1787/9789264209138-en.

OECD (2013), *Providing Agri-environmental Public Goods through Collective Action*, OECD Publishing, Paris, http://dx.doi.org/10.1787/9789264197213-en.

OECD (2012a), *A Framework for Financing Water Resources Management, OECD Studies on Water*, OECD Publishing, Paris, http://dx.doi.org/10.1787/9789264179820-en.

OECD (2012b), *Water Quality and Agriculture: Meeting the Policy Challenge*, OECD Publishing, Paris. http://dx.doi.org/10.1787/9789264168060-en.

OECD (2008), *Costs of Inaction with Respect to Air and Water Pollution, in: Costs of Inaction on Key Environmental Challenges*, OECD Publishing, Paris, http://dx.doi.org/10.1787/9789264045828-4-en.

OECD (2006), *Cost-Benefit Analysis and the Environment: Recent developments*, OECD Publishing, Paris. http://dx.doi.org/10.1787/9789264010055-en.

OECD (2005), *Reducing water pollution and improving natural resource management*, in: Sustainable Development in OECD Countries: Getting Policies Right, OECD Publishing, Paris, http://dx.doi.org/10.1787/9789264016958-en.

Olmstead, S. (2010), "The Economics of Water Quality". *Review of Environmental Economics and Policy*, Vol. 4, No. 1, pp. 44–62.

ONS (2016), *Natural Capital: overview of the ONS-Defra Natural Capital project and all related publications*, www.ons.gov.uk/economy/nationalaccounts/uksectoraccounts/methodologies/naturalcapital [accessed 04/06/2016].

Parris, K. (2012), Impacts of agriculture on water pollution in OECD countries: Recent trends and future prospects, *In: Water Quality Management: Present Situations, Challenges and Future Perspectives*, Eds. A.K. Biswas, C. Tortajada and R. Izquierdo, Routledge, Oxfordshire, UK.

Pimentel, D. (2005), *Environmental* and economic costs of the application of pesticides primarily in the United States, *Environment, Development and Sustainability*, Vol. 7, pp. 229–252.

Randall, A. (1983), The problem of market failure, *Natural Resources Journal*, Vol. 23, pp.131-148.

Rothamsted Research (2005), Cost effective diffuse pollution mitigation, Final Project Report, commissioned by DEFRA, UK.

Searchinger, et al. (2008), Use of U.S. croplands for biofuels increases greenhouse gases through emissions from land-use change, *Science*, Vol. 319, pp. 1238–1240.

Secchi, S., et al. (2011), Potential water quality changes due to corn expansion in the Upper Mississippi River Basin, *Ecological Applications*, Vol. 21, No. 4, pp. 1068–1084.

Shortle, J.S. and R.D. Horan (2013), Policy Instruments for Water Quality Protection, *Annual Review of Resource Economics*, Vol. 5, pp. 111-138.

Shortle, J. S. and R.D. Horan (2001), The economics of nonpoint pollution control, *Journal of Economic Surveys*, Vol. 15 (3), pp. 255 – 289.

Shortle, J.S. et al. (2012), Reforming agricultural nonpoint pollution policy in an increasingly budget-constrained environment, *Environmental Science and Technology*, Vol. 46, pp 1316-1325.

Shortle, J. and T. Uetake (2015), "Public Goods and Externalities: Agri-environmental Policy Measures in the United States", OECD Food, Agriculture and Fisheries Papers, No. 84, OECD Publishing, Paris, http://dx.doi.org/10.1787/5js08hwhg8mw-en.

Smith, L.E.D. and K.S. Porter (2010), "Management of Catchments for the Protection of Water Resources: Drawing on the New York City Watershed Experience", Regional Environmental Change 10(4).

Sobota, et al. (2015), Cost of reactive nitrogen release from human activities to the environment in the United States, *Environmental Research Letters*, Vol. 10 (2), pp. 14.

US EPA (2015), A compilation of cost data associated with the impacts and control of nutrient pollution, Office of Water, United States Environmental Protection Agency, EPA 820-F-15-096.

US EPA (2006), "National Primary Drinking Water Regulations: Long Term 2 Enhanced Surface Water Treatment Rule; Final Rule", Federal Register, Vol. 71, No. 3, *Rules and Regulations*, 5 January.

US EPA (2001), Arsenic in Drinking Water Rule – Economic Analysis, Report No. 815-R-00-026, Washington D.C.

Van Grinsven, et al. (2013), Costs and Benefits of Nitrogen for Europe and Implications for Mitigation, *Environmental Science and Technology*, Vol. 47 (8), pp 3571–3579.

Van Grinsven, H., Rabl, A. and T. de Kok. (2010), Estimation of incidence and social cost of colon cancer due to nitrate in drinking water in the EU: a tentative cost-benefit assessment, *Environmental Health*, Vol. 9 (58).

Water UK (2013), CAP Reform: A future for farming and water, Water UK, London, United Kingdom.

Webster-Brown, J. (2015), '*Call for Cantabs to think about future of water*', Lincoln University Press Release, 13 April 2015. www.scoop.co.nz/stories/ED1504/S00028/call-for-cantabs-to-think-about-future-of-water.htm [accessed 28/04/2016].

WWAP (World Water Assessment Programme) (2012), *The United Nations World Water Development Report 4: Managing Water under Uncertainty and Risk*, UNESCO, Paris.

Diffuse Pollution, Degraded Waters
Emerging Policy Solutions
© OECD 2017

Chapter 3

Emerging policy instruments for the control of diffuse source water pollution

This chapter examines innovative policy approaches to help meet the challenge of diffuse pollution. It presents and draws lessons from a select number of case studies submitted by OECD member countries and discussed at the OECD Workshop on Innovative Policy Responses to Water Quality Management held in March 2016. All case studies are provided in full at www.oecd.org/water.

Key messages

The complexity of water quality challenges in a rapidly changing world means that new, locally-adapted and innovative solutions are often required. Policy makers seek innovations that help meet water quality objectives at a lower cost; current water pollution control policies could often deliver better environmental outcomes at a lower cost for society. In particular, policy innovations that realise greater diffuse pollution control are essential for advances in water quality.

Policy approaches used to date for the control of diffuse pollution tend to be voluntary and developed at the local level via partnership, around watersheds or cities, and most often include government's paying farmers to reduce pollution. However, there is evidence that voluntary participation may not reach the major polluters and subsidy-based programmes can have limited impact due to public budget constraints and a lack of environmental regulations on diffuse pollution.

Enforceable limits on diffuse pollution are required and can be designed in relation to an acceptable level of risk to water quality. The greatest challenge of regulating outputs of diffuse pollution is how to allocate a pollution "cap" (or maximum permitted load) to individual land owners in a way that is equitable and cost-efficient. A natural capital based approach to allocating diffuse pollution limits is an emerging development that has the ability to reach the full economic potential of natural resources, based on the underlying capacity of the soil to filter and retain water and nutrients.

Payments for ecosystem services can generate financial flows, and engage stakeholders in water management, but must be underpinned by enforced environmental regulations so as to achieve additionality. Water quality trading offers a way to promote practices that reduce pollution at least cost to society by revealing preferences and water pollution costs. Markets and payments for ecosystem services are not a perfect substitute for robust regulations. Equity and fairness in burden sharing do not preclude efficiency.

It is necessary, and more effective, to use a combination of the available policy mechanisms, including regulatory, economic and voluntary regimes, to improve pollution control. Stakeholder engagement through inclusive water governance is increasingly recognised as critical to secure support for reforms, raise awareness about water risks and costs, increase users' willingness to pay, and to handle conflicts. Striving for consensus through collaborative governance can pave the way for action provided safeguards to enable participation and equal representation of all stakeholders are provided for.

Introduction

There is a regulatory imbalance on the control of point source versus diffuse source pollution. Policy instruments to address point source and diffuse source pollution are different due to their different characteristics[1]. The dominant approach to managing point source pollution is regulation via effluent standards (OECD, 2004), which has created significant improvements in water quality in OECD countries (OECD, 2013a). Such point sources of pollution are under the direct control of polluting firms and utilities and can be accurately monitored, policed and enforced at a reasonable cost. For example, the US Clean Water Act established a regulatory framework with stringent federal effluent standards for industrial and municipal point source pollution that led to significant water quality improvements in the US (Shortle and Horan, 2013), although there is some debate over whether incremental costs have exceeded the incremental benefits since the late 1980s (as discussed in chapter 2).

For diffuse source pollution, the most prominent methods are the use of voluntary payments and regulatory instruments. However, voluntary measures (e.g. payments to farmers to incentivise pollution reductions) have generally had limited success (Shortle and Horan, 2013), and regulatory measures to control diffuse water pollution are typically poorly enforced (Parris, 2012). There is evidence that voluntary participation may not reach the major polluters (OECD, 2012, 2004) and subsidy-based programmes can have limited impact due to public budget constraints (Shortle and Horan, 2013). Indeed voluntary codes of good practice have not proved able to remedy a problem that does not stem from a lack of information, but from the absence of internalisation of pollution costs (OECD, 2004). There is opportunity to better design, target and implement voluntary and regulatory instruments to achieve water quality outcomes.

Economic instruments, such as pollution taxes, charges and water quality trading, could be strengthened and used more extensively to increase the cost effectiveness of pollution control and promote innovation. Taxes on inputs (i.e. product charges and other proxies for pollution) can be used to create incentives to reduce pollution from urban and rural sources (Hoffmann and Boyd, 2006), and to raise funds for investment in water quality infrastructure and management (Köhler, 2001; Hoffmann et al., 2006). Taxes on impervious surfaces in urban areas can incentivise reductions in stormwater runoff and finance a greater proportion of urban land to be connected to a drainage system with stormwater treatment.

In essence, alternative approaches to monitor and manage diffuse pollution are required. The following sections will present new and innovative approaches to control, finance and govern water pollution, with a particular focus on diffuse source pollution. While the chapter is split into subsections depending on the type of policy instrument, in reality a policy mix is required for effective outcomes. For example, information instruments are required for determining the maximum nutrient load a catchment can sustain before certain water quality indicators are breached; stakeholder engagement is necessary to determine the acceptable risks and desired water outcomes at the local level; regulatory nutrient pollution allowances are required for diffuse pollution before markets can be established to achieve economic efficiency; and compensations may be required to improve equity outcomes and stakeholder agreement.

Regulatory approaches

Regulatory policy instruments to reduce diffuse pollution from agriculture typically restrict the use of polluting inputs such as fertilisers, manure and pesticides, and require farm management practices that reduce pollutants reaching water bodies. This is for two reasons: *i)* there good evidence to show that "good" or "best" management practices can reduce the losses of diffuse contaminants, and *ii)* up until recently, there was limited ability to adequately calculate farm scale losses of diffuse source contaminants with computer models. As an example, the following two case studies present regulations to reduce diffuse nitrogen pollution from agriculture through mandatory best management practices in the EU (Box 3.1) and Canada (Box 3.2).

Box 3.1. **European Union nitrate pollution prevention regulations**

The Nitrates Directive (1991) aims to protect water quality across Europe by preventing nitrates from agricultural sources polluting ground and surface waters and by promoting the use of good farming practices. The Nitrates Directive complements the Water Framework Directive and is one of the key instruments in the protection of waters against agricultural pressures. The first step in implementation of the Directive is the identification of waters that are polluted or could become polluted if no action is taken. These are considered as waters where the concentration of nitrates is above 50 mg/L or that could contain (if no action is taken to reverse the trend) more than 50 mg/L of nitrates and waters that are eutrophic or could become eutrophic if no action is taken. Eutrophication caused by phosphorus as well as nitrates must be considered when designating nitrate vulnerable zones (NVZs). Member states can also choose to apply measures to their whole territory rather than designating specific zones. Farmers within NVZs must comply with specific measures such as:

- Limiting when nitrogen fertilisers can be applied on land in order to target application to periods when crops require nitrogen and prevent nutrient losses to waters;
- Limiting the conditions for fertiliser application (on steeply sloping ground, frozen or snow covered ground, near water courses, etc.) to prevent nitrate losses from leaching and run-off;
- Requirements for a minimum storage capacity for livestock manure;
- Crop rotations, soil winter cover, and catch crops to prevent nitrate leaching and run-off during wet seasons; and
- Limits on the total amount of livestock manure that may be applied to land.

Every four years member states are required to report on: *i)* nitrates concentrations in groundwaters and surface waters; *ii)* eutrophication of surface waters; *iii)* assessment of the impact of action programme(s) on water quality and agricultural practices; *iv)* revision of NVZs and action programme(s); and *v)* an estimation of future trends in water quality.

Source: Case study provided by the EU.

Box 3.2. **Agricultural nutrient management regulations, Canadian provinces**

In response to ongoing problems with eutrophication and algal blooms, Canadian provinces have mandated nutrient management plans at the farm level through regulatory changes for some time. For example, buffer strips around surface water and groundwater sources have become a common requirement to limit nutrient leaching. Federal programmes to reduce diffuse nutrient pollution – Environmental Farm Plans and the Environmental Stewardship Incentive - are designed and implemented at provincial level, which enables policy to be adapted to local circumstances and facilitates the transfer of knowledge.

Box 3.2. Agricultural nutrient management regulations, Canadian provinces *(cont.)*

Under municipal by-laws, the location of manure storage, as well as setback distances from neighbouring properties or streams, may be regulated. Examples of regulatory measures to reduce diffuse pollution from agriculture at the provincial level include:

- Ontario: The Nutrient Management Act (2002) sets out regulatory requirements for certain nutrient management practices and requires farmers to document these practices to reduce risk of water contamination by agricultural sources. The practices regulated include the management of manure (e.g. storage and application), application of non-agricultural materials (e.g. sewage bio-solids and vegetable processing wastes) and the treatment of manure and other materials in on-farm anaerobic digesters.
- Manitoba: The Livestock Manure Mortalities Management Regulation (1998) prescribes various requirements for the use, management and storage of livestock manure to reduce water pollution from livestock. Permits are required for the construction, modification or expansion of manure storage facilities and specific constraints, such as maximum livestock population, fencing restrictions, restrictions to drainage and water work, apply on crown land.
- Quebec: The Agricultural Operations Regulation (2002) seeks to address the problem of diffuse pollution caused by agricultural activity, by achieving an effective balance of phosphorous in the soil to maintain soil fertility and limit losses from excessive use of manure. It includes norms for livestock buildings and manure management, and restrictions on land use to limit water pollution. Other regulations deal with the use of fertilisers and pesticides in agriculture.

Source: OECD (2015a).

In Israel, national policy calls for the gradual replacement of freshwater allocations for agriculture with reclaimed effluents in an effort to improve water security, water quality and nutrient recycling. The case study from Israel (Box 3.3) details the number of regulations that have increased the cost-effectiveness of, and investment in, tertiary wastewater treatment plants which is enabling unrestricted irrigation of crops while ensuring diffuse pollution is limited. Similarly, Turkey is exploring options for the safe reuse of treated water and wastewater in agriculture, industry, tourism, energy and households.

Box 3.3. Regulations to incentivise wastewater reuse and reduce water pollution, Israel

For Israel, overcoming the challenges of an arid climate and scarce natural water reserves has always been a vital necessity for the growth of the population and economy since the founding of the state. Currently, Israel annually requires almost a billion cubic metres per year (MCM/yr) more water than average natural replenishment provides. Nevertheless, average annual sustainable natural water consumption has been achieved, providing for all of the country's water needs, via innovations in wastewater reuse and desalination.

National policy calls for the gradual replacement of freshwater allocations to agriculture by reclaimed effluents. A number of important advances in policy were required for sustainable wastewater reuse to become a reality while safeguarding groundwater from diffuse pollution:

1. In 1992, the Ministry of Health (which controls potable water quality and manages the effluent irrigation permitting system) released new regulations that set secondary wastewater treatment plant (WWTP) quality standards for biochemical oxygen demand and total suspended solids. As a result, municipalities built intensive WWTPs with national loans of USD1.5 billion and restricted wastewater reuse was permitted for agricultural irrigation.

Box 3.3. **Regulations to incentivise wastewater reuse and reduce water pollution, Israel** *(cont.)*

2. In 2001, the Water and Sewerage Corporations Law was passed. The Law transferred the management and ownership of water and sewerage infrastructure and services from public municipalities to corporate entities. This was the first step in the transformation of the water sector from administrative management to a more commercial orientation. The process was initially a voluntary one, but since 2008 state loans in the water sector are given only to private water and sewerage corporations.
3. In 2006, Parliament approved the establishment of a National Water Authority (under the Ministry of National Infrastructures, Energy and Water Resources) with overall responsibility for water, sewage and water resources management policy. One of the main principles introduced was that water tariffs should enable full cost recovery in the water sector, including costs of water conveyance, piping systems and wastewater treatment. This principle, together with construction of desalination plants along the Mediterranean coast, dramatically increased water prices in Israel for all sectors. Today, the domestic sector is paying about USD 2.6 per cubic meter of potable water. Therefore, the incentive for wastewater reuse was further enhanced.
4. In 2010, tertiary WWTP water quality standards came into force to enable unlimited irrigation of crops while ensuring diffuse pollution is limited and water and soil quality are protected. The standards have 37 water quality parameters, including heavy metals, nutrients and oxygen demand.

Investments in WWTPs and the volume of treated wastewater have increased over 5-fold and 3-fold respectively between 1992 and 2008. The reuse of wastewater treated at the secondary level for use as irrigation (under restrictions) was originally encouraged through zero fees from the WWTP. Farmers and water associations had the responsibility to construct the piping and distribution systems, and the reservoirs to collect the water for irrigation. Since wastewater quality was upgraded to tertiary level, farmers pay USD 0.3¢ /m^3; the additional cost to attain the improved water quality standard (from secondary to tertiary level treatment) is approximately USD 0.1¢/m^3 (capital + O&M). The benefits of tertiary treatment and unlimited effluent irrigation are much greater than the costs. In addition, freshwater ecosystems are flourishing due to a permanent and steady environmental flow of treated wastewater that can be used for various purposes downstream of WWTPs. This has triggered further investment in advanced treatment technologies at WWTP's, carrying out nitrogen and phosphorus reduction, filtration and disinfection.

Currently, treated wastewater constitutes about 21% of total water consumption in Israel and approximately 45% of agricultural consumption. Out of a total of about 510 million cubic metres of wastewater produced in Israel annually, 97% of the wastewater is collected and about 85% of it is reused. 52% is treated to tertiary level and 41% is treated to secondary level. The ultimate objective is to treat 100% of Israel's wastewater to a level enabling unrestricted irrigation in accordance with soil sensitivity and without risk of pollution to soil and water sources.

Sources: Summary of case study provided in full by Alon Zask, Ministry of Environmental Protection, and Adi Yefet, Israel NewTech - National Energy & Water Program, Israel; Rejwan and Yaacoby (2015).

Regulating diffuse pollution outputs

Instead of regulating the use of inputs to reduce diffuse source pollution, policy makers are increasingly looking to computer models to regulate calculated outputs of pollution from individual landowners and "cap" various users to meet calculated total maximum pollution loads for catchments. Advances in computer modelling can predict diffuse pollution based on farm practices (such as crop rotations, stocking ratios, tillage practices, fertiliser and pesticide applications, irrigation), and the hydrological, soil, and geographical conditions that effect the transport of pollutants to surface and groundwater bodies (Fishmana et al., 2012). Land managers can innovate farm and land management practices within their pollution cap without being restricted by the inputs they use. Furthermore, regulating pollution through proxies such as fertiliser use and livestock numbers can be less effective at reducing pollution[2] (OECD, 2010). Models can then be progressively adapted to account for innovations that improve water quality.

There are some limitations to modelling, one of which is simplification and assumptions made of complex environmental systems. Models are also only as good as the data with which they are calibrated and validated. Uncertainties in data and model components can propagate to other model components and model outputs. Greater data collection and reliability can lead to continual improvements in the accuracy of models, and their capacity to make informed policy decisions. A concerted scientific effort can reduce uncertainties in predicting water quality and the consequence and benefits of current and alternative trends and scenarios. Some regions of New Zealand have used nutrient modelling to inform policy design (Box 3.4). Water quality modelling is an important part of the Total Maximum Daily Load Management System in Korea, which aims to control both point and diffuse pollution sources through a permitting system (see case study below).

Box 3.4. Nutrient modelling in New Zealand

OVERSEER®, a national model for farm-scale nutrient budgeting and loss estimation, calculates nutrient flows in a productive farming system and identifies risks of environmental impacts through nutrient loss, including run-off and leaching. The model was originally developed as a tool for farming to create nutrient budgets and has been adapted to overcome barriers that arise from an inability to clearly identify diffuse source polluters. It is recognised as the best tool currently available for estimating nitrate leaching losses from the root zone across the diversity and complexity of farming systems in New Zealand. A summary of the model inputs and outputs are summarised in the table below.

Inputs: Farm level	Inputs: Management block level (i.e. paddock/field scale)	Outputs
Farm location Types of blocks and block areas (e.g. pastoral, fodder crop, house, scrub, wetland, riparian) Types of enterprises (e.g. pastoral, cropping) Stock Stock numbers, breed Production Placement (grazing off, wintering pads) Types of structures Effluent management of structure Stock management on structure Type of effluent management system Supplements imported and where they are fed Wetlands	Topography Climate Soil type Drainage Soil fertility tests Pasture type Supplements made on the block Fertiliser applied Irrigation applied Effluent applied Animals (type, timing) grazing the block Crop rotation; crops grown – yield, fertiliser applied, harvesting method	*Nutrient budget.* Nitrogen (N) sources: atmospheric, fertiliser, animal transfer, supplements fed on block, irrigation and nutrients out N losses: produce (e.g. milk), animal transfer, supplements (e.g. hay), leaching/runoff, atmospheric (e.g. N_2O). *Farm-level and block-level reports.* e.g. Total N lost to water for blocks and farm; Average N concentration in drainage based on N leached; N surplus per block. *Advisory reports.* e.g. N conversion efficiency (%); total GHG emissions; maintenance fertiliser requirements.

Box 3.4. Nutrient modelling in New Zealand *(cont.)*

OVERSEER® can, and has, supported environmental policy development, most notably around Lake Taupō and as part of Horizons One Plan in the Manawatū-Wānganui region. New Zealand farmers will increasingly use the model to develop nutrient management plans and budgets, as required by regional councils. While such a model is essential for enabling a water pollution cap to be imposed, it is accepted by both farmers and regional councils that it has high uncertainties. The model is not designed to provide economic analysis, so outputs need to be combined with other economic models to assess the impacts of options on the farm business.

The accuracy of OVERSEER® will be critical to maintaining the credibility of policies. In order to improve the accuracy of OVERSEER®, further investment is required to better calibrate and validate the model under different soil types, farm types, farm management practices and mitigation methods, and under extreme weather conditions (such as high rainfall), uncommon situations (e.g. specialist types of horticulture) and under highly complex operations. Improved versions of OVERSEER® will need to be recognised in policy so that innovative mitigation methods can be implemented by farmers.

Source: OECD (2017).

Case study: The Total Maximum Daily Load Management System, Korea[3]

Rapid economic growth in Korea has increased the demand for water tremendously, and accelerated urbanisation and agriculture intensification has caused deterioration of water quality in rivers and lakes, which has resulted in an increase of social conflicts regarding water use among various stakeholders.

In 2004, the total maximum daily load (TMDL) programme was introduced to improve water quality management policy, which had previously focused on the regulation of the concentration of point sources of pollution. The TMDL allocates pollution load reductions necessary to reduce the sources of pollution and achieve desired water quality. The TMDL program in Korea is aimed at water quality improvement and economic growth simultaneously. Since implementation of the TMDL, further reductions in point source pollution have been achieved. However, the proportion of diffuse sources in relation to point sources has increased; the proportion of total pollution attributed to diffuse pollution is projected to reach over 70% by 2020. In response to the shifting challenge, the control of diffuse sources of pollution has been the focus of water policy since 2011. Central government oversees water quality management by local government and sets overarching policies for both economic development and environmental protection. Water quality targets are set periodically for each of the four main watersheds (the Nakdong River, the Geum River, the Youngsan Seomjin River and the Han River). Local water quality targets and implementation plans are then established to achieve the overarching target for the watershed.

Target parameters are BOD and total phosphorus. It is envisaged that the TMDL system will go beyond simple parameters like BOD and total phosphorus, to also manage non-biodegradable substances in the future. Targets near the boundaries between provinces and cities are required to be notified so that water quality targets can be attained in co-operation.

Permissible total maximum daily pollutant loads are calculated using scientific water quality modelling at the watershed, local and individual property levels. Economic development, population growth, pollutant reduction and local development planning are considered together. The TMDL management system clarifies the responsibility of each relevant entity by identifying each pollution load by local government, sub-local government and individual polluter, with a view to meeting and staying on the water quality target.

Once the water quality targets are set, governors and mayors develop detailed local development plans and the annual plans for pollution reduction, with a view to meeting the load allocation of each watershed. In co-operation with stakeholders, governors and mayors then decide how to allocate pollution load permits to individuals in order to attain and maintain the overarching target for the watershed. It is up to each local government and the stakeholder how to allocate these pollution loads. Technical Guidelines for TMDL Management provided by the National Institute of Environmental Research require pollution load permits be set through water quality modelling, considering equity, efficiency and effectiveness of reducing pollution loads. Several methods are given as options in the Technical Guidelines, such as the grandparenting approach, catchment average approach, sector average approach, proportional to revenue generation, and the minimum cost method, among others (see Table 3.1 for an explanation of these methods). Voluntary allocation through stakeholder co-operation and engagement is also encouraged in the Technical Guidelines (NIER, 2014).

Implementation and performance of the TMDL system is evaluated every year by central government. When improvements are required, the central government may ask a governor or mayor to establish and take necessary measures: for example, putting further restrictions on urban and industrial development projects, suspension or cutbacks of financial support, or restriction on installation or modification of facilities where discharging wastewater. Central government also offer support for implementing the TMDL system. Since 2004, when TMDL system was implemented, government subsidies have been provided to support investments in wastewater treatment plants, and land purchases to retire sensitive areas from intensive land use (such as riparian buffer strips) with the aim of meeting water quality targets. In addition, for keeping the TMDL system sustainable, the development of new pollution reduction technologies and approaches through R&D projects is a factor attributable to the success of the TMDL system.

The nationwide introduction of TMDL management system is considered to have contributed to improving water quality, particularly in areas where the existing water quality was worse than the water quality target. In 2013, water quality targets were achieved for 81% of rivers. Figure 3.1 is a demonstration of the success of the TMDL system in the Nakdong, Geum and Yeongsan-Seomjin Rivers. However, water quality targets were achieved in only 12% of lakes; Korean's lakes and reservoirs are particularly vulnerable due to the high residence time in comparison to rivers (as are most lakes around the world). Strict management is considered necessary to achieve continuous water quality improvements in future stages of TMDL implementation (Kim et al., 2016).

Figure 3.1. **Water quality improvements under the Total Maximum Daily Load Management System, Korea**

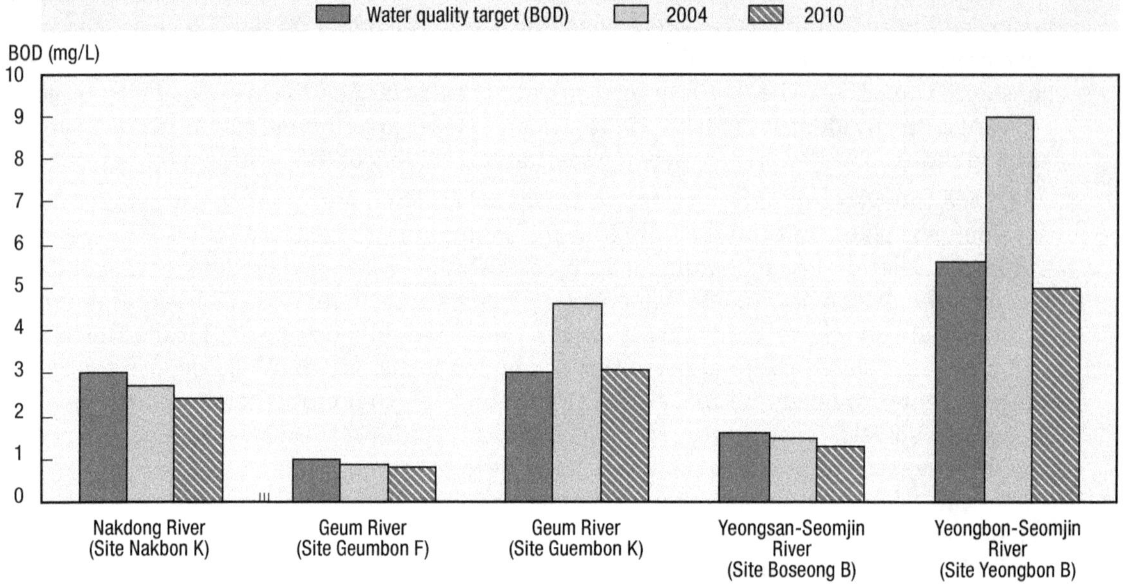

Source: Summary of case study provided in full by Jihyung Park and Sang-Cheol Park, Ministry of Environment, Korea.

Allocating diffuse pollution allowances within a cap

The greatest challenge of regulating outputs of diffuse pollution is how to allocate pollution "caps" (or maximum permitted loads) to individual land owners in a way that is equitable and cost-efficient. Table 3.1 describes the different approaches for allocating modelled nutrient discharge allowances to achieve water quality objectives within a catchment. Nutrient management measures must then be put in place to avoid increases in nutrient output from land use changes or intensification if a catchment is at full pollution allocation.

Table 3.1. **Allocation approaches for nutrient discharge allowances**

Allocation approach	Description
Grandparent	Nutrient discharge allowances (NDAs) allocated based on nitrogen leaching rates during a baseline or benchmarking period and proportional to nutrient reduction target.
Catchment average	All landowners are given the same NDA regardless of land use (i.e. the average of total nitrogen discharge from land-based sources per productive land area).
Land cover average	Landowners managing a specific land cover (e.g. pasture, forest, arable) are given the same NDA per productive land area.
Sector average	Landowners within the same sector (e.g. dairy, sheep, beef, horticulture) are given the same NDA per productive land area.
Natural capital	NDAs allocated based on the biophysical potential of the land, soil and environment. Land use capability can be used as a proxy for natural capital, with a greater NDA allocated to higher class land. The allocation is independent of existing land use.
Nutrient vulnerability	NDAs allocated based on the nutrient leaching capacity of the soil. A greater NDA would be allocated to land with lower "vulnerability" or greater capacity for filtering nutrients. The allocation is independent of existing land use.
Least cost	NDAs allocated to meet nutrient cap/target at lowest cost.
Auction	Auction or reverse auctions to allocate NDAs. Those who can afford to pollute or wish to intensify production can buy the rights to do so. In facilitating such an auction, pollution rights can shift to the most productive land users.
Community-negotiated allocations	NCAs negotiated and allocated among stakeholders.
Ballots	Lucky draw.
Merit-based criteria	Based on best economic, environmental and/or social returns.

Note: NDA is nutrient discharge allowance.
Source: Adapted from Daigneault et al. (2017).

The efficiency and equity of each of the allocation approaches differ based on their existing land use, land characteristics, stakeholder preference, and the stringency of the regulation. A study by Daigneault et al. (2017) demonstrates that there is no most or least preferred allocation option based on cost-efficiency criteria. Allocation of pollution allowances should be made at a level that is consistent with good land management practice to encourage reductions in pollution, particularly those who lag behind such practices. Additional regulations may be necessary to protect vulnerable areas such as near water bodies (e.g. requiring riparian zones be planted and fenced off from livestock).

In terms of equity, the grandparent approach (the most frequently used approach), can be considered inequitable with historic polluters rewarded, which may also be the polluters most likely to be able to reduce pollution at a lower cost. The grandparent also has high opportunity costs for those property owners who have not developed land or who may wish to intensify production. The natural capital approach is an emerging approach that decouples the pollution allocation from land use. Such an approach encourages the adaptation of land activity to better use soil and water resources, and shift land use activities to more sustainable outcomes based on the underlying natural capital stocks of soils. The approach is being implemented in two of New Zealand's sixteen regions (see case study below). Economic analysis and stakeholder engagement can identify the most efficient and equitable allocation option in a given context.

Case study: The natural capital approach to allocating diffuse nitrogen pollution, Manawatu-Wanganui, New Zealand[4]

Freshwater is the backbone of New Zealand's economy. In addition to tourism, recreation, power generation and cultural identity, it is vital to the primary sector. New Zealand is unique among developed countries as nearly three quarters of its export earnings are generated from the primary industries of agriculture, horticulture, viticulture, forestry, fishing and mining. The expansion of intensive dairy farming, in particular, has yielded significant economic benefits, but has also had significant consequences for water quality (OECD, 2015b).

In recognition of the need for limits on natural resource allocation, the National Policy Statement for Freshwater Management (2014) directs all Regional Councils to set limits on water quality and quantity in all water bodies above a specified minimum level required to ensure ecosystem and human health by 2025. The Statement also requires that overall water quality within a region must be maintained or improved. For the agricultural sector, systematic setting of water quality and water quantity limits in catchments throughout New Zealand poses both opportunities and risks.

Regional Councils have taken different approaches to address the issue of setting nutrient loss limits. For the Horizons Regional Council, who manages the natural resources of the Manawatu-Wanganui region, an innovative approach using the natural capital of the soil has been used to allocate nitrogen diffuse pollution limits to individual farmers as part of the One Plan[5].

Allocating a nutrient loss limit based on the natural capital of the soil (inherent capability) in a catchment offers an approach for developing policy that is linked directly to the type of underlying natural biophysical resources in the catchment, recognising that different soils and topographies differ in not only their productive potential, but also their ability to filter and retain water and nutrients under the pressure of grazing animals (Mackay, 2009).

The New Zealand Land Use Capability Survey (Lynn et al., 2009) is a land use classification system that been used in New Zealand to help achieve sustainable land development and management on individual farms, in whole catchments, and at the district, region, and the national level since 1952. The Land Use Capability (LUC) system has two key components. Firstly, Land Resource Inventory (LRI) is compiled as an assessment of physical factors considered to be critical for long-term land use and management, and secondly, the inventory is used for LUC Classification, whereby land is categorised into eight classes according to its long-term capability to sustain one or more productive uses. The LRI is compiled from a field assessment of five factors: rock type, soil type (including depth, structure, texture, drainage, and nutrient supply), slope angle, erosion type and severity, and vegetation cover. The LRI is supplemented with information on climate, flood risk, erosion history and the effects of past land management practices. The greater the limitations imposed by the physical constraints of the soil and the environment, the greater the number and complexity of corrective practices that are required for productive use.

The productivity indices (i.e. attainable potential carrying capacity) in the LUC survey offer a proxy for the natural capital of soils. An attraction of the approach is that this information is already established as the basis for land development and evaluation, and it is available throughout New Zealand. Using the productivity indices, combined with the national OVERSEER® nutrient budget model[6] (Box 3.4), nitrogen losses can be calculated based on the LUC classification. As the limitations to use of the soil increase (i.e. Class 1 to 7), the underlying capacity of soil to sustain a legume-based pasture system declines, as does the potential nitrogen loss by leaching, since carrying capacity also decreases. In other words, the most vulnerable landscapes that have little natural capital and/or lack versatility in either land use options and/or mitigation strategies should have the most stringent limits on nitrogen leaching.

The nitrogen leaching limits for each LUC in the Operative One Plan (Table 3.2) are staged over a period of 20 years (see table below). The nitrogen leaching limits have legal effect ranging from 1 July 2014 to 1 July 2016, depending on the Water Management Zone. Landowners with existing and new intensive land use activities (dairy farming, commercial vegetable growing, cropping and intensive sheep and beef farming) must prepare and implement a nutrient management plan. Nutrient management plans outline specifically how farming operations will comply within the One Plan nitrogen leaching limits, thereby allowing for a customised farm scale level assessment and plan to deliver on the water quality outcomes that is reflective of each farms unique collection of biophysical resources and business structure and operation. In practice, it provides a vehicle for the progressive implementation of actions that might first target adoption of low cost strategies that achieve a high cost-benefit return, progressively adopting more expensive input cost options, and constraints on production as farms strive to achieve nitrogen and other contaminant targets. As new knowledge and technologies become available, they can be readily incorporated into the nutrient management plans.

Table 3.2. **Cumulative nitrogen leaching maximum by Land Use Capability Class, Manawatu-Wanganui Region, New Zealand**

Period (from the year that the rule has legal effect)	LUC I	LUC II	LUC III	LUC IV	LUC V	LUC VI	LUC VII	LUC VIII
Year 1	30	27	24	18	16	15	8	2
Year 5	27	25	21	16	13	10	6	2
Year 10	26	22	19	14	13	10	6	2
Year 20	25	21	18	13	12	10	6	2

Source: Horizons Regional Council (2014).

Horizons Regional Council began the One Plan development process with community consultation in 2004 and first notified the plan in May 2007. There has been strong opposition by farmers to many of the changes, in particular to those concerned with the cost of nutrient management regulating unsustainable practices. The majority of the Plan was settled through mediation, but 20 % of appeals went to the Environment Court to rule on some of the more contentious issues surrounding nutrient management. The appeal process of the One Plan has caused long delays for the Horizons Regional Council and users; after some amendments to the original proposed One Plan, the One Plan became operative on 19 December 2014 (over 7 years since notification). Pathways forward are required to manage the challenges associated with the longer-term transition to such a policy, in a similar way to which stranded assets in the climate sector need to be managed. Bank lending to the dairy sector, in particular, should be monitored to ensure that these policy requirements are being taken into account when assessing the ability of farmers to repay debt (OECD, 2014).

It is still early days in the One Plan's implementation, and to date its impact on surface water quality outcomes in the priority catchments have not been assessed. Striking the balance between achieving the environmental gains set out within the One Plan, with the expectations of communities seeing tangible improvements in surface water quality, with the rate at which farmers could or should be required to comply with this environmental legislation, is still being openly debated in the Region.

The Natural Capital approach sets limits on emissions to water that are linked directly to the underlying land resources, by providing a boundary condition that future land uses and practices have to operate within to achieve the required water quality outcomes. This approach encourages the adaptation of land activity to better use soil and water resources, and shift land use activities to more sustainable outcomes. It also encourages transitioning from policy that looks to regulate land use to one that ensures the finite underlying resources are being sustained and available for future uses. The approach may not always optimise economic outputs within water quality limits, particularly in the short term as the full transition is made to shift intensive land use from low class soils to high class soils. Trading of water pollution allowances could assist with such a transition. Shifting to a natural capital based approach offers a basis for assessing the capability of wider landscapes to provide multiple ecosystem services for a range of desired outcomes beyond just economic growth and water quality. The approach has more recently been adopted in the Hawke's Bay region[7].

Policy can be designed to further improve the economic efficiency of nutrient allocation approaches. For example, water quality trading (discussed in the following section) offers a solution whereby farmers within a catchment can exchange nutrient allocations to achieve a cumulative nutrient limit for a catchment in a flexible manner that maximises economic efficiency and maintains environmental integrity. In order to improve equity, those most affected by allocation regimes can be compensated, where necessary. Where trading may not be appropriate, communities and individuals can use outcomes-driven approaches and collaborative management to plan mitigations. The following sections will discuss a range of these approaches.

Economic instruments

Growing recognition of the importance of diffuse pollution problems has stimulated interest in the design of economic instruments to control pollution sources (Shortle and Horan, 2001), especially since fiscal consolidation and budgetary constraints at national and local level has reduced financial assistance (subsidies) to reduce pollution (Shortle et al., 2012).

Pollution charges and taxes

Putting a price on negative externalities is one way negative impacts to the environment can be internalised, and economic valuation can assist policy makers to make informed decisions on the most beneficial solutions for society. Pollution taxes, user fees and product charges can be used to create incentives to reduce pollution from urban and rural sources, increase the cost-effectiveness of pollution control and promote innovation in pollution control strategies (Hoffmann et al., 2006).

Examples of water pollution taxes do exist, but on the whole, the use of economic instruments in water pollution control is much less common than in air pollution control (Hanley et al., 2013) and in the control of diffuse source pollution than in point source pollution. The heterogeneous impacts and damage costs of water pollution makes their management more difficult than air pollution. Additional reasons for the slow uptake of economic instruments in the management of water pollution may include: political resistance from polluters; limited data on the costs of environmental degradation; difficulties in measuring diffuse sources of pollution and attributing them to landowners; and the complexities of ambient pollution concentrations which are a function of both point and diffuse pollution sources, natural background levels, watershed characteristics, fate and transport parameters, and stochastic environmental variables (Figure 1.2, Chapter 1) (Shortle and Horan, 2001).

Taxes on polluting inputs (i.e. product charges on inputs that are believed to have environmentally harmful effects) and pollution charges (i.e. a charge based on the quantity of pollutants that are discharged into the environment) can be used to create incentives to reduce pollution from urban and rural sources (Hoffmann and Boyd, 2006), and to raise funds for investment in water quality infrastructure and management (Köhler, 2001; Hoffmann et al., 2006). There is a large variation in how and for which pollutants water pollution taxes and charges are implemented in different countries or regions. Table 3.3 provides a few examples.

Table 3.3. **Examples of features of pollution charges in selected OECD countries**

Country	Levied by	Tax name	Specific tax base	Tax structure
Australia	State	Water effluent charge	Volume, pollution content (types of pollutants)	per kg assessable load
Canada	Province	Charge on discharge	Volume and pollution content	per litre or per tonne
Denmark		Diffuse source	Chemical deterrents of insects and mammals	tax on retail price
France		Diffuse source	Pesticides	per kg
		Water effluent charges	Households	per m^3
Netherlands		Tax on the pollution of surface waters	BOD, COD and heavy metals, for large polluters	per pollution unit
Sweden	Municipality	Wastewater user charges	Wastewater and drinking water	varies by municipality; full cost charging
		Diffuse source	Pesticides	per whole kg active constituent

Source: OECD database on Policy Instruments for the Environment (Accessed 20/03/2016).

In France, taxes are used to encourage and finance reductions in water pollution and water consumption (Box 3.5). In Denmark, a pesticide tax has been applied since 2013 so that farmers are taxed according to the environmental and health toxicity of pesticides used rather than their nominal value. In Norway, since 1999 the pesticides tax has been area-based with seven tax bands according to the environmental and health related risks of the pesticides. The tax was initially introduced in 1988 as a revenue raising tool, but was revised in 1999 to reflect a stronger objective of reducing the use of pesticides. This system has been effective in encouraging more conservative use of pesticides and provides an incentive to use less harmful products (OECD, 2010; Withana et al. 2014). Although revenue raised is often earmarked for improvements in water quality, revenue should be allocated to the general budget of governments for use for policy and projects that may render the greatest benefit to society.

Advances in nutrient pollution modelling (e.g. Box 3.4) provides an opportunity to tax diffuse pollution outputs, rather than taxing inputs as proxies such as fertiliser use and livestock numbers, which can be less effective at reducing pollution[2] (OECD, 2010). Using such models, pollution charges could be directly proportional to the amount of pollution generated.

Box 3.5. Taxes to encourage and finance reductions in water pollution and water consumption in France

French water policy is based on using water and pollution taxation to finance actions to protect and restore water resources and aquatic environments, and to follow the polluter-pays and user-pays principles. This system, implemented by the Water Agencies, involves stakeholders at basin level working together in Basin Committees to determine the size of charges to levy within statutory national limits. This participatory model facilitates the acceptance of taxes by liable entities and permits periodic adjustment in order to take new issues into account, remain representative of water users and retain its levels of acceptability.

Taxes are designed to internalise environmental externalities in the price and pollution of water resources, taking into consideration: *i)* costs and impacts generated by pollutants released into water bodies; *ii)* costs and impacts of obstacles in rivers, such as dams and weirs which can effect environmental flows, freshwater ecosystems and water quality; and *iii)* costs and impacts relative to a chronic shortage of a water resource, generating conflicts of use and changes in flows which have an impact on the management of water quantity and water quality (as flow reduction causes a greater concentration of pollution). However, given the low tax levels used (see table below), it is unlikely the external environmental and opportunity costs are fully internalised.

Tax revenues are earmarked for reinvestment in water quality and scarcity improvements at the basin level and are managed by the Water Agencies. National government has capped total tax revenue from water-related taxes for the 2013-2018 period at EUR 13.8 billion. The allocation of tax revenue to specific projects is assessed on environmental effectiveness and economic efficiency. Records of funded projects are made publically available to comply with transparency obligations.

2013 water-related tax levels, Rhone-Mediterranean and Corsica Water Agency, France

Water agency tax	Uses	Calculation	2013 tax level
Water abstraction charge	All users	Proportional to water withdrawn	Depends on the use, the level of water scarcity, and the collective or non-collective management of water
Domestic pollution	Urban users	Proportional to water consumption	EUR 0.23/m^3
Industrial pollution	Industrial users	Proportional to generated pollution	Depends on type of pollutants
Sewer systems modernisation	Users connected to a public sewerage network	Proportional to volume discharged in sewer network	EUR 0.15/m^3
Diffuse pollution from livestock	Farmers with >90 livestock units	Proportional to livestock unit, which factors type of livestock and age	EUR 3.0 per livestock unit from the 41st unit
Hydroelectricity production	Hydroelectric operators (>1 billion cubic metres per year diverted)	Proportional to volume of diverted water	EUR 1.2/billion m^3 and per metre of waterfall height
Obstacles in rivers	Those who modify natural river systems (except hydroelectric operators)	Proportional to length of the barrier	EUR 150/m
Storage in low water level periods	Entities who store water	Proportional to the volume stored in low water level and high demand periods	EUR 0.01/m^3 stored
Protection of freshwater environments	Recreational fishers	Per recreational fisher	EUR 8.8/year/adult plus EUR20.0 for specific fish species

Sources: Summary of case study provided in full by Maude Jolly and Emmanuel Steinmann, Ministry of Ecology, France; Montginoul et al. (2015).

Water quality trading

It is recognised that there can be inefficiencies associated with regulations, taxes and subsidies. Where property rights are established, market-based instruments, such as water quality trading[8], involves less control from government and offers a mechanism for achieving a cost-effective allocation of environmental effort across alternative sources, without environmental regulators knowing the abatement costs of individual agents (Parris, 2012). When carefully designed, trading can address water quality problems by using trade ratios that define allowable rates of exchange between sources that reflect the marginal damage of emissions from each source (Farrow et al., 2005; Muller and Mendelsohn, 2009).

Although they can be complex in nature, water quality markets can stimulate innovation and often achieve water quality targets at a lower social cost than traditional performance standards, taxes and payments/subsidies (OECD, 2013b). Water quality trading can potentially enable continued growth in a capped watershed without jeopardising water quality, and water quality goals may be met at a faster pace than without trading. To date, market-based instruments to address water pollution in OECD countries have been limited (primarily to point-point sources), but there is growing interest in their use. The case of point-diffuse source water quality trading to reduce nutrient pollution of Chesapeake Bay, United States is illustrated as an example below. Another case study, the Lake Taupō nitrogen market, New Zealand is the first diffuse source pollution market in the world, enabled by the nutrient model OVERSEER® (Box 3.4) to cap nitrogen emissions at the catchment scale and allocate discharge allowances to individual farmers for trading (OECD, 2015c; 2017).

Case study: Lessons learnt from water quality trading, Chesapeake Bay, United States[9]

The Chesapeake Bay watershed on the east coast of the United States is considered a "national treasure and resource of worldwide significance" (Chesapeake Bay Restoration Act of 2000). It is also the largest estuary in the United States and one of the largest and most productive in the world. However, the Chesapeake Bay watershed has suffered from excess nutrients and sediment for decades. In particular, diffuse pollution from agriculture, has been largely unregulated and remains a significant contributor to water quality impairments not only in the Chesapeake Bay watershed, but across the country. The Chesapeake Bay Foundation (2014) has estimated that the Chesapeake Bay area provides more than USD 107 billion in ecosystem services, such as flood and hurricane protection, air and water purification, and food production every year. However, if current trends continue, it is estimated that the region could lose USD 5.6 billion annually in benefits.

After voluntary attempts at improving water quality failed to deliver adequate results, the United States Environmental Protection Agency (EPA) worked with the states in the watershed in 2010 to pioneer the largest and most complex total maximum daily load (TMDL) for nutrients and sediment in the country. Maximum loading rates were determined by the Chesapeake Bay Watershed Water Quality Model (calibrated to in-stream monitoring data) to ensure protection of the Bay ecosystem and its tidal tributaries. In total, the TMDL called for a 25% reduction in nitrogen, 24% reduction in phosphorus, and 20% reduction in sediment. In consultation with the states, the EPA allocated nutrient reduction requirements to the states and the major river basins based on their pollution contribution (i.e. a grandparenting approach). Other factors considered included the relative effectiveness of nutrient reduction efforts among river basins, and the degree to which the loads were controllable. Pollution control mechanisms to meet these allocations, and fully restore the Bay, must be in place by 2025. In addition to the reduction requirements, all new sources of pollution loads must be offset, as there are no allocations for future growth.

After each major basin had an allocation, the states sub-allocated their major river-basin loads to individual sources in their Watershed Implementation Plans. The decisions guiding this process varied by state, but generally, point sources of pollution face stringent nutrient discharge limits and are expected to achieve limit of technology. Diffuse pollution from agriculture remains largely unregulated but is collectively subject to a load allocation under the TMDL. States report to the EPA each year on what nutrient mitigation practices and activities have been implemented. This gets fed into the Chesapeake Bay Watershed Water Quality Model which produces a "progress run", using the latest in-stream monitoring data, to determine the annual loading rates, reductions and improvements in water quality.

Traditionally, TMDLs and other water quality goals have been primarily addressed through traditional and costly command-and-control approaches on point sources. However, as shown in Figure 3.2, costs for nitrogen mitigation varies greatly among sectors, with agricultural diffuse source practices having significantly lower costs than point sources. Based on the dramatic price differentials among sectors for nutrient mitigation options, water quality trading has emerged as a market-based mechanism for cost-effectively meeting water quality goals and the TMDL in three states – Pennsylvania, Virginia and Maryland. Furthermore, given that there are no load allocations for future growth under the Chesapeake Bay TMDL, there is the risk of stifling development without flexible mechanisms in place for handling growth. Trading is considered to be a critical mechanism for accommodating growth through acquiring pollution offsets.

Figure 3.2. **Cost comparison of options to reduce a nitrogen loading, Chesapeake Bay Watershed, United States**

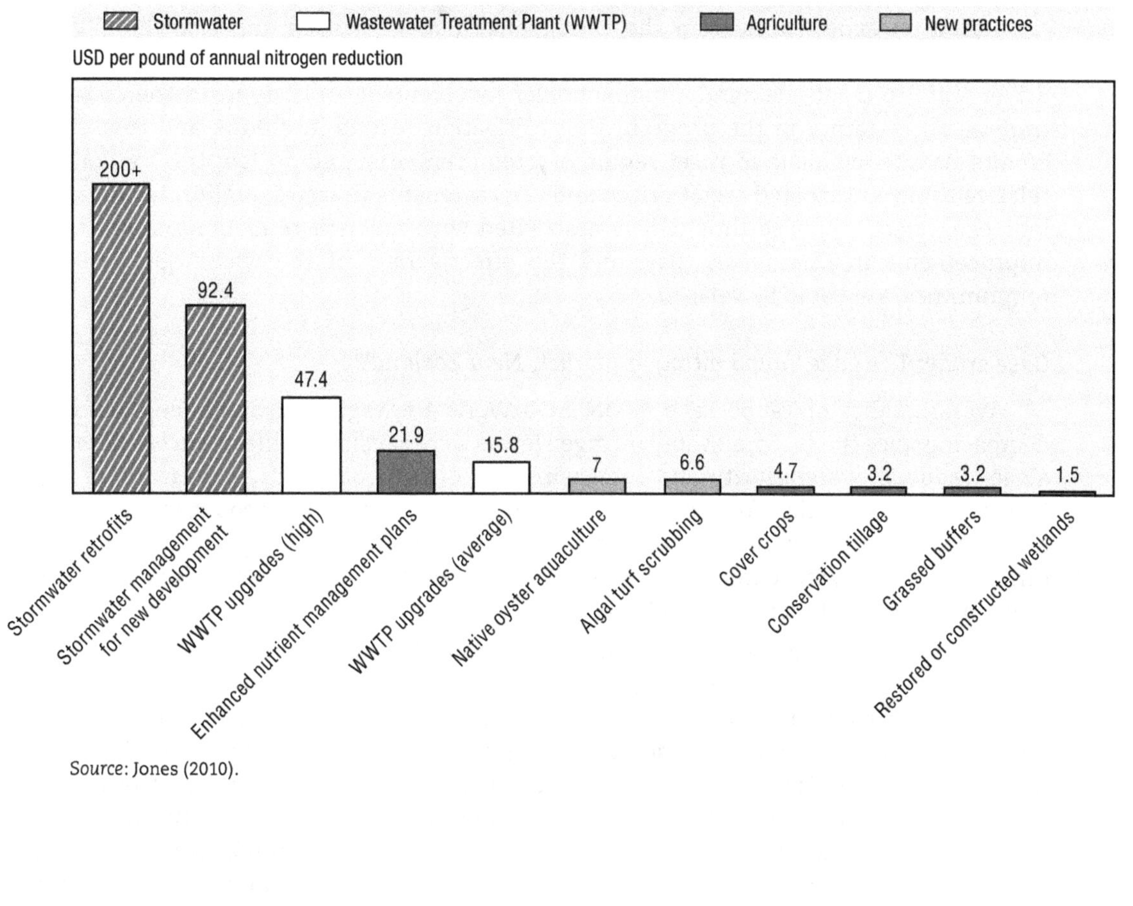

Source: Jones (2010).

The regulations are slightly different in each state, but in general, trading is permitted between regulated point source entities, such as wastewater treatment plants, urban stormwater municipalities and concentrated animal feeding operations (point-point trading), and with diffuse pollution sources (point-diffuse trading). The buyer is therefore the regulated point source entity, and the seller may be another regulated point source entity or an unregulated entity, such as a farmer who has gained credits for reducing diffuse pollution. Regulated entities and new and expanding facilities can offset their pollution loads by purchasing credits from other point sources or from diffuse sources at lower cost that could be attained on site. Water quality credits are issued to entities that have achieved pollution reductions beyond their allocated pollution load (after onsite verification).

Water quality trading in the Chesapeake Bay watershed has enabled regulated entities to meet permit requirements at a reduced cost than under traditional command and control approaches, and credit generators, such as farmers, have earned additional revenue through the sale of credits. For example, in Pennsylvania, trading has been successful as a flexible compliance option for WWTPs. From 2013-2015, 600 000 - 1.1 million nitrogen credits, and 55 000 – 100 000 phosphorus credits have been sold annually. These represent a mix of point-to-point trades and point-to-diffuse trades. Virginia's Nutrient Credit Exchange has proven to be an effective mechanism for facilitating compliance trading between point sources with permanent phosphorus offsets selling for upwards of USD 20 000/pound, largely generated by land conversion activities. Maryland's nutrient trading program has yet to experience any trades. This is likely due to a lack of binding regulations, which may have led to uncertainty and risk for regulated entities to purchase credits in lieu of addressing requirements onsite.

Based on experiences from the Chesapeake Bay watershed and elsewhere, critical requisites for a water quality trading programme to be viable include: *i)* a strong regulatory driver (cap) to create demand; *ii)* stakeholder involvement and buy-in to the concept of trading; *iii)* certainty in the program that mitigation results in credits and that enough credits will be available to meet regulatory requirements; and *iv)* low transaction costs relative to the anticipated credit prices and improvements in water quality. Because of the long ecosystem response-time delays associated with nutrient reductions, water quality improvements in Chesapeake Bay, and the effectiveness of the water quality trading programmes, are yet to be verified.

Case study: The Lake Taupō nitrogen market, New Zealand[10]

Water quality of Taupō Lake, a UNESCO World Heritage Site, had been consistently decreasing since the 1970s; elevated nitrogen levels were causing proliferation of microscopic algae, reducing water clarity and increasing the growth of weeds in near shore areas. Diffuse source pollution from pastoral farming was estimated to account for over 90% of anthropogenic nitrogen inflows to Lake Taupō, despite efforts of Taupō farmers to reduce diffuse pollution with extensive stream fencing, planting and riparian land retirement under a Taupō Catchment Control Scheme in the 1970s.

In response, the government, Waikato Regional Council, Taupō District Council and Ngati Tuwharetoa (the local iwi) implemented an innovative diffuse water quality trading project, comprising three components: *i)* a cap on nitrogen emission levels within the Lake Taupō catchment by OVERSEER® (see Box 3.4 for description); *ii)* establishment of the Taupō nitrogen market; and *iii)* formation of the Lake Taupō Protection Trust to fund the initiative. The costs were to be spread across local, regional and national communities; the independent Lake Taupō Protection Trust was established in 2007 to use public funds (NZD 79.2 million) to buy back allocated nitrogen allowances to retire land and to reduce the economic and social impacts of the nitrogen cap. The trading scheme was also complemented by the New Zealand Emissions Trading Scheme, which came into force during the early stages of the project and advanced the achievement of nitrogen reductions; the promotion of land-use change from pasture to forestry not only surrendered nitrogen discharge allocations, but also received carbon sequestration credits for a time.

The target was to reduce manageable nitrogen emissions to 20% below current recorded levels, so as to restore water quality and clarity to 2001 levels by 2080. The reduction of manageable nitrogen was initially estimated at 153 tonnes of nitrogen but later increased to 170.3 tonnes annual discharge reduction by 2018 as a result of improved benchmarking data. This was equivalent to reducing 153 tonnes of nitrogen annual discharge by 2018. Based on this catchment cap, each farm was allocated an individually-calculated nitrogen discharge allowance, consistent with the desired reduction in emission levels. This permitted them to leach a certain level of nitrogen every year, based on their previous levels of nitrogen use. This approach, known as "grandparenting" (see Table 3.1 for explanation), was not without contention among different stakeholders. Forest landholders and sheep and beef farmers saw it as inequitable; land development had opportunity costs, and farmers who had been a major cause of the pollution of Lake Taupō were rewarded with higher allowances. The OVERSEER® model provided the basis for generating farm-specific figures to establish nitrogen discharge allowances.

The ability to trade through establishment of the Taupō nitrogen market was a critical part of the negotiations. Farmers wanted flexibility and ability to increase production, or to receive direct financial benefits for reducing nutrient leaching. As part of the market design, only landowners in the catchment can buy, sell and trade nitrogen allowances; this was thought necessary to avoid outside investors purchasing and trading allowances for capital gain. The cap-and-trade policy began in July 2011. By 2013, all farms in the catchment had applied for resource consents and had been benchmarked for their nitrogen discharge allocation. By mid-2015, the Trust had secured contracts to meet the 170.3 tonnes of nitrogen target reduction, and there had been 12 private nitrogen discharge allowance trades between regulated farmers (totalling 18 tonnes of nitrogen).

A recent review of the Lake Taupō nitrogen market (Duhon et al., 2015) found that a cap on nitrogen has limited the nitrogen leaving agricultural land. However, the cap has also had negative impacts on those affected, including reduced ability to intensify production, decreased land values and significantly increased administration and compliance costs. All of these trade-offs were necessary to address the environmental problem of excessive pollution. The Lake Taupō Protection Trust, which funded decreases in nitrogen, significantly reduced the costs borne by farmers but came at a high cost to government. Motu (2015) suggests that regulators should continue to reduce trading transaction costs. Making allowance price information available to farmers would be useful, as would any policies that increase the future liquidity of the market.

The policy package has been fully implemented. It is providing the flexibility for land to move to its highest value and best use, and still meet the overall nitrogen load reduction targets. The use of the model OVERSEER® is essential to the cap-and-trade programme, providing incentives for farmers to reduce nitrogen emissions. The Lake Taupō Protection Trust has permanently retired 20% of the original nitrogen discharge allowances. New lower-nitrogen ventures are emerging in the catchment, such as growing olives, farming dairy sheep, and producing and marketing "sustainable" beef. The environmental certainty enables development of added-value products with credible green branding. It also generated positive environmental impacts, particularly carbon sequestration, from the reforestation of more than 5 000 ha of land to pine plantations.

Payment for ecosystems services

Payment for ecosystems services (PES) can internalise water pollution and other environmental externalities through the Beneficiary Pays Principle by incentivising polluters to change their behaviour. For example, downstream beneficiaries of improved water quality (such as water utilities, industry, city councils and recreational users) pay upstream farmers in return for land management practices that reduce pollution. Payments to reduce diffuse water pollution from agriculture in OECD countries are most commonly for a reduction or cease in the use of fertilisers and pesticides, for the retirement of arable land, and for the establishment of riparian buffer strips.

In Germany, a voluntary payment scheme offered by the municipal water provider of Munich gives payment to upstream farmers to convert to organic farming processes to reduce nitrates and pesticides (Box 2.4). The scheme was more environmentally effective and cost-efficient than upgrades in water treatment to remove nutrients and pesticides.

In England, PES schemes are gaining in popularity with water utilities, with improved outcomes not only for water quality and reduced water treatment costs, but also for biodiversity, flood management and environmental flows (Box 3.7). To reduce concerns about equity that can arise if PES payments are seen to "reward polluters" while neglecting producers already demonstrating best practice, there is a need for collective compliance by farmers with baseline regulation so as to achieve "additionality" in response to PES incentives (OECD, 2013b).

Box 3.6. Co-operation with farmers for catchment protection in Munich, Germany

The Mangfall Valley in the Bavarian Alps supplies around 80% of Munich's drinking water for its 1.2 million inhabitants. The Valley is predominantly used by farmers and agricultural producers whose activities were causing slow but significant increases in nitrates (15 mg/L) and pesticide concentrations (0.065 µg/L) in the city's water resources. To address the issue, in 1991 the municipal water provider Stadtwerke München (SWM) implemented a voluntary payment scheme to encourage local farmers to adopt more sustainable organic farming practices.

After estimating the target area using hydro-geological models, SWM launched a public information campaign targeting 120 farmers (mainly dairy producers). The payments were constructed to cover the expected lost income and investments needed to switch to organic farming; more precisely, farmers received a payment of EUR 280 per hectare per year (ha/year) for the first 6 years after the change, and EUR 250/ha/year for the following 12 years.

The programme successfully halved nitrate concentration to 7 mg/l; the price increase for final urban consumers due to the payment scheme (EUR 0.005 per cubic metre [EUR/m^3]) was lower than the avoided cost of water-treatment facilities (EUR 0.23/m^3). More than 90% of the farmers adhered to the programme. The Munich area is now considered the largest and most active market for organic farming products in Germany.

One of the key success factors of the programme was the city's strong involvement in purchasing and promoting the organic products from the Mangfall Valley. Not only did the city purchase the organic farming goods to supply its schools and municipal restaurants, it also funded several marketing and advertising campaigns aimed at creating a brand identity for the targeted area's agricultural goods. These measures helped build trust between urban and rural water consumers and – together with a clearly defined set of legal rules regulating organic farming practices in Germany – reduced the contractual and transaction costs of implementing the payment scheme.

Sources: OECD (2015d); Grolleau and McCann (2012).

Box 3.7. Collaboration with farmers and Payment for Ecosystem Services schemes in England

Problems of water pollution from point sources such as factories and other industrial activity have declined through both structural change in England's economy and effective regulation. Although some legacy water quality problems from industrialisation (e.g. old mine workings, now managed through public investment in the absence of historic polluters), and morphological alteration to waterbodies as a result of human activity (e.g. navigation, hydropower, flood defence activity), the most significant modern water quality problem is diffuse pollution, particularly from agriculture. Agricultural subsidy frameworks, inconsistent land use planning systems, and under-reported and under-regulated diffuse pollution, have contributed to water quality pressures.

The primary pollutants which water utilities have to deal with are nitrates, phosphates, sediments and pesticides. It is estimated that since the 1989 privatisation of the water sector (supervised by three regulators and national government), water utilities have invested around GBP 1.7 billion in traditional drinking water treatment approaches to reduce the levels of pesticides and nitrates. The scale of these costs has been a key driver for the industry to pursue new ways of working with land managers to reduce pollution at the catchment scale. In recognition that, in a wider social sense, it is not efficient to pollute at source through sub-optimal land management practices and then have to consume resources downstream to remove pollution, water utilities began considering diverting investment from traditional water treatment into land management as payment for ecosystem services (PES).

"Upstream Thinking" is South West Water utility's catchment management scheme which has been applying natural landscape-scale solutions to water quality issues since 2008. The PES scheme draws upon the knowledge and expertise of a number of partners including South West Water, the Devon Wildlife Trust, the Cornwall Wildlife Trust, the Westcountry Rivers Trust, the Exmoor National Park Authority, and local farmers to improve raw water quality at source. Over the 2015-20 period, the latest GDP 11.8 million programme is focussing on 11 catchments across Devon and Cornwall. The target for the programme is 750 farms and 1 300 ha of moorland and other semi-natural land under revised management.

Upstream Thinking targets priority pollutants associated with different catchments – typically nutrients, pesticides, and sediments. Farm advisers visit farms and carry out an assessment resulting in a whole-farm plan to reduce nutrients, pesticides and sediments. This includes a water management plan and future capital investment proposals targeted at water quality improvements. Up to 50% of capital investment proposals are funded by Upstream Thinking. These can include improvements to slurry storage, fencing to keep livestock out of rivers, providing alternative water sources for livestock, and improved pesticide management including investment in new equipment such as weed wipers which deliver targeted doses of herbicide.

The Upstream Thinking programme has also successfully investigated and restored over 2 000 hectares of sensitive upstream land on Exmoor in 2010-15 to improve peatland, and reduce sediment loads and flood risk downstream. The overall programme is fully endorsed by the Environment Agency, Natural England and the Drinking Water Inspectorate. The work is targeted to benefit 15 water treatment works supplying 72% of the total daily water to customers.

Although physical evidence is emerging on the water quality benefits of working with land managers in catchments through water companies making investments to pay for ecosystems services, the economic evidence on the costs and benefits of the approach has been slower to emerge. In its 2011 report, *From Catchment to Customer*, Ofwat (the economic regulator) acknowledged a lack of hard economic evidence on the net benefits of land management PES approaches. It also highlighted the role for polluter-pays mechanisms alongside the beneficiary-pays approach which characterises the water company schemes.

Box 3.7. Collaboration with farmers and Payment for Ecosystem Services schemes in England *(cont.)*

Nevertheless, Ofwat does see a role for PES schemes, saying "Water customers could legitimately expect to pay for those elements of catchment management that bring direct and measurable benefits to them, under the principle of paying for ecosystem services" (Ofwat, 2011). In 2009, Ofwat approved ultilities' proposals to spend £60m on water quality investigations and PES schemes throughout England and Wales, representing something of a departure for Ofwat.

Sources: Summary of full case study provided by Nick Haigh, Defra, UK; UK Environment Agency (2015); National Audit Office (2010); Ofwat (2011).

Financing mechanisms

Improved financing models and public-private partnerships (PPPs) are required to finance investments to replace aging infrastructure, and adapt to population growth, urbanisation, climate change and increasing water quality regulation. For example, it is estimated that the United States faces a water infrastructure investment deficit of USD 384 billion over the next 20 years (US EPA, 2011). Capital is needed for investment in thousands of miles of pipes, as well as thousands of treatment plants, storage tanks, and other key assets to ensure the public health, security, and economic well-being of cities, towns, and communities (US EPA, 2011).

Two case studies are presented:

- The United States State Revolving Funds provides an example of a sustainable infrastructure financing model. Set up with "seed money" from the United States Congress, the State Revolving Funds capitalise a state-administered financial assistance program to build and upgrade wastewater treatment plants and drinking water infrastructure, as well as invest in other projects to improve water quality (such as measures to reduce diffuse pollution and water recycling). In doing so, the Funds support a longer transition and ample flexibility to set up long-term financing to promote state and local self-sufficiency.
- London has established a novel Government Support Package, to attract private financiers and reduce insurance liabilities, to deliver the Thames Tideway Tunnel project – a major construction undertaking to intercept London's combined sewer overflows for treatment to improve water quality of the River Thames.

Case study: State revolving funds for water quality protection, United States[11]

In 1987, after the United States Congress amended the Federal Water Pollution Control Act, the federal government handed down responsibility for funding and constructing wastewater infrastructure, and improving the nation's water quality, to state and local government. In doing so, the United States Congress created the Environmental Protection Agency-administered, state-domiciled Clean Water State Revolving Fund (the "CWSRF" or "SRF") programme to replace the EPA administered Construction Grants Program. This programme was designed to replace direct federal assistance that had been provided in the form of grants, in favour of a new funding model that offered states resources to operate a financial assistance program on behalf of local governments where dedicated resources would revolve in perpetuity.

There were a number of objectives served by the change to the CWSRF. Firstly, it conformed to a long held federal view that local governments are ultimately responsible for funding projects needed to protect the nation's water quality. Secondly, shifting day-to-day control for financial assistance to the states was thought to be a better delivery model for state and local governments. The change was also expected to reduce the claim on federal resources supporting municipally-owned treatment works, and to expand the menu of eligible projects, including projects that address diffuse sources of pollution.

The CWSRF programme created authorised federal funding ("seed money") for a limited five year period. Federal funding was contingent on a 20% state match for every federal dollar appropriated. The law provided that funds could support a number of financial assistance options including loans, the purchase of debt obligations, to use as pledged security for municipal bond transactions, financial guarantees and investment. Repayments of obligations to the state by eligible recipients of CWSRF financial assistance would provide for build-up of a renewable source of capital for future investments. The intention was that states would have flexibility to set priorities and administer funding.

Since its creation, states have used the CWSRF financial assistance to support more than USD 110 billion in financial assistance. Such assistance has largely been delivered as loans, debt obligation purchases and bond security. Through 2015, financial assistance has leveraged federal investment by 280%. The early success of the CWSRF encouraged the United States Congress to create the Drinking Water SRF (the "DWSRF") as part of the 1996 amendments to the Safe Drinking Water Act. The reasons for doing so were: *i)* increased nutrient contamination (from both point and diffuse sources) was putting drinking water sources at risk, *ii)* increased regulation of a number of new contaminants required large investments in treatment technology to meet statutory requirements, and *iii)* many of the nation's 52 000 small community water systems were likely to lack the financial capacity to meet the rising costs of compliance with Safe Drinking Water Act. Through 2015, the DWSRF had leveraged federal investment by 171% and delivered USD 30 billion in financial assistance to public water systems. The majority of states have leveraged federal and state investment by entering the bond market to boost programme capacity.

The table below illustrates the amount of funds leveraged from federal seed money of the SRFs between 2010 and 2015. Bond financings undertaken on behalf of multiple local governments have merited triple-A ratings due to the level of equity overcollateralization and loan portfolio diversification. Under most SRF credit structures, the net loan rate to the borrowers is a function of the over-collateralized cash flows. This is, in turn, a function of federal tax law which limits investment to the related tax-exempt bond issues' cost of funds.

Table 3.4. **Clean Water and Drinking Water State Revolving Funds: Funds Available for Projects**

2010-2015 (Billions of U.S. Dollars)

Funding mechanism	CWSRF	DWSRF
Federal Capitalization Grant	39.5	17.5
State Match	7.4	3.2
Net Leveraged Bonds	35.7	8.1
Net Interest Earnings	8.2	1.7
Net Transfers between SRFs	-0.3	
Funds for re-funding	-1.6	
Administration Set Aside	-1.6	-0.6
Other Set Asides		-2.1
Total funds	**87.3**	**27.8**

Note: CWSRF = Clean Water State Revolving Fund; DWSRF = Drinking Water State Revolving Fund
Source: Environmental Protection Agency, United States

The CWSRF and the DWSRF have enabled States to shape their programmes to their specific needs, characteristics and capabilities. Projects served under the CWSRF include wastewater treatment plant (WWTP) upgrades to secondary and tertiary treatment, combined sewer overflow corrections, measures to reduce diffuse pollution, and recycled water. In addition to investment in water infrastructure, the DWSRF was critical in quickly channelling much needed resources to boost economic activity and remediate storm-damaged infrastructure in the aftermath of Hurricane Sandy in 2012.

Twenty five years in, the attributes of the SRFs increasingly look like that of an endowment, designed to operate in perpetuity. The CWSRF and the DWSRF are currently in a position to not only meet their mission objectives as presently understood, but are also in a position to expand their product offerings to support new water quality and public health funding solutions while employing endowment investment strategies to grow the capital base. For these opportunities to be realised, EPA and the SRF administrators need to reconsider existing legal and institutional frameworks that present barriers to higher programme performance. This can be done by *i)* implementing endowment-like investment strategies that could subsequently accelerate returns based on general market investment; and *ii)* developing targeted investment strategies that could yield both investment returns and innovative technological advancement (such as investment in green infrastructure and market-based solutions to diffuse pollution). The CWSRF and the DWSRF also require a more concerted effort to engage and educate stakeholders of the SRFs' true capabilities.

Case study: A public-private approach to delivering the Thames Tideway Tunnel, United Kingdom[12]

London's combined sewer system was constructed over the period 1859-75. Designed to serve a maximum of 4 million people, and to accommodate both domestic sewage and stormwater runoff, the significant increase in London's population to 8 million over the last 150 years means that there is no longer sufficient capacity in the sewer system during periods of rainfall. As a consequence, the combined sewer overflows (CSOs) originally designed as "safety valves" discharge untreated sewage and stormwater to the River Thames, on average, once per week (in excess of 40 million tonnes of sewage each year).

After assessing a range of options to address the problem, the most economically feasible solution was to build a new tunnel – the Thames Tideway Tunnel (TTT) – under the bed of the Thames, to intercept the CSOs, transferring wastewater to the Beckton sewage treatment works for processing. Other options considered, such as widespread introduction of sustainable urban drainage systems to deal with surface water runoff, and constructing a new separate drainage network for stormwater water (in both cases, to release pressure on the main combined network), were found to be too costly or not feasible. End-of-pipe, in-river solutions to deal with CSO pollution once it had entered the Thames, such as skimmer vessels and oxygen bubbling, were also considered, and although not found satisfactory as a long-term solution, have been employed on a limited scale to mitigate the worst impacts pending the construction of the TTT.

An economic cost-benefit analysis of the TTT estimates welfare benefits of GBP 7.4-12.7 billion for a whole-life project cost of GBP 4.1 billion (Defra, 2015; Eftec, 2015). However, despite the economic benefits significantly out-weighing the capital investment costs, the exceptionally high risk profile of the project (i.e. the unprecedented scale and major tunnelling work under one of the world's most complex cities) inflated the cost of borrowing (high risk premium) and reduced the ability of Thames Water (the private water utility) to secure finance for the project through the normal capital markets at something approaching a reasonable cost for its customers.

In response to the difficulty of attracting finance for the TTT, the national government, Ofwat (the financial regulator) and Thames Water put in place public-private arrangement to enable risks to be managed and underwritten to the point that private capital markets were prepared to finance the project on reasonable terms. The key principle behind developing a risk-sharing model for the TTT was that different parties have different capacities to mitigate and absorb different risks, which in turn affects financing costs and ultimately the feasibility of a project.

A novel Government Support Package for the TTT project was constructed, which provided contingent public financial support under very specific circumstances. It is provided for a fee and comprises five agreements:

- *Supplemental compensation agreement:* the government provides an insurance facility to the project for cover above the limits of commercial insurance, including if commercial insurance were available at the beginning of the project but subsequently becomes unavailable.
- *Contingent equity support agreement:* if the costs of completing the TTT are forecast to escalate beyond a specified point (the "Threshold Outturn") the project will have the option of requesting that government make an injection of equity to allow it to be completed. If the government receives such a request, it is committed to provide this equity, subject to its right to discontinue the Project (see below).
- *Market disruption facility:* in the event that the project is unable to access debt capital markets as a result of a sustained period of disruption in these markets, the government would provide temporary liquidity.
- *Special administration offer agreement:* if the provider should go into Special Administration and not have exited after 18 months, the government commits to either make an offer to purchase the provider (at a price at its discretion), or to discontinue.
- *Discontinuation agreement:* the government will have the right to discontinue in a number of circumstances, in particular, in the event that the costs of completing the TTT are predicted to escalate beyond the Threshold Outturn and the project requests an injection of equity from the government, or if insurance claims exceed a specified amount. Where the government opts to discontinue the project, it commits to paying compensation to existing equity and debt investors. This agreement of the government support package acts to ensure that the government does not assume unlimited liabilities.

As a result of the public-private risk-sharing model and the Government Support Package, an infrastructure provider and a construction consortium have now both been appointed and construction of the TTT is proceeding and expected for completion in 2023. The competition for both the infrastructure provider and the construction contracts were highly competitive. The winning bid for the infrastructure provider offered a weighted average cost of capital of 2.497%, which is fixed, subject to the terms of the project licence, until the first price review following construction. The construction procurements delivered a target build cost which is unchanged from that estimated in 2011 (GBP 4.1 billion).

Under a scenario without the support of the government, it is very likely that the TTT project would have been unfeasible; assuming financing would be forthcoming, an estimate of the cost was approximately GBP 10 billion. Therefore, as a result of the public-private risk-sharing model, water and wastewater bills for households (the ultimate funders of water sector investment) are kept at reasonable levels, and protected in the event of catastrophic risks (the maximum additional cost is estimated at GBP 20-25 per customer per annum by the mid-2020s (in 2015 prices), down from an earlier estimate of a maximum of GBP 70-80 (in 2011 prices).

Costs to the government have been carefully defined to only arise in very specific, low probability circumstances. In the absence of risks materialising, the cost to the government is zero, but the presence of the Government Support Package as a "backstop" gives comfort to the private sector and enables efficient private financing.

A collective management approach to water quality management

Stakeholder engagement through inclusive water governance is increasingly recognised as critical to secure support for reforms, raise awareness about water risks and costs, increase users' willingness to pay, and to handle conflicts (OECD, 2015c). In the last decade, stakeholder engagement has gained traction in the water sector in OECD member countries in response to new legislation and guidance requiring greater inclusiveness, transparency and accountability (OECD, 2015c). A collective management approach with stakeholder engagement in setting regulations at the local level (with an overarching national water quality baseline) can create buy-in, increase trust in government processes, and ultimately find effective solutions to achieve desired water quality outcomes. This is demonstrated in the case of the Canterbury region, New Zealand, which also includes challenges and requisites for success. An additional case study on the "Catchment Based Approach" in England further illustrates the importance of local governance arrangements for improving water quality management (see oecd.org/water for the case study).

Case study: A collaborative governance model: The Canterbury Water Management Strategy, New Zealand[13]

The region of Canterbury contains 70% of New Zealand's irrigated land, 65% of the nation's hydroelectricity storage capacity, an extensive groundwater system, highly prized coastal lagoons, lowland waterways valued for cultural and recreational use, and world-renowned braided river systems. In the early to mid-2000's, public concerns about deteriorating water quality, reduced reliability of water supply for irrigation, and dissatisfaction with the adversarial approach to water management became widespread. The community at large had reached breaking point due to over-allocated water resources, pressure from droughts, and degraded water quality, primarily from diffuse pollution from agriculture.

In response to these problems, the Canterbury Water Management Strategy (CWMS) was developed over the years 2007-2009 to provide a framework for a collaborative approach to water management and with targets across all interests in water. The CWMS is a new paradigm for water management in Canterbury. It has three key features: *i)* delivering environmental, economic, cultural and social outcomes together ("parallel development", defined as 10 targets); *ii)* A shift from effects-based management of individual resource consents for individual landowners to integrated management of catchments; and *iii)* A collaborative governance framework where "local people, plan locally".

The CWMS is a partnership between the regional council, district councils, and the Māori (indigenous) tribal authority in Canterbury - Ngāi Tahu. The region is divided into 10 zones (based on a combination of hydrology, administrative boundaries, and communities of interest) and each zone has a Zone Committee comprising of four-eight representatives of the local community with a range of interests in water, representatives from district and regional councils, and representatives from the local rūnanga (Māori sub-tribe). The task of each committee is to develop Zone Implementation Programmes with recommendations and actions for achieving each of the ten targets laid out in the CWMS.

The committees operate at the "collaborate" step on the engagement staircase (Figure 3.3), and "involve" their local communities in their deliberations and decision-making (in the form of field trips, workshops, and one-on-one meetings) to ensure that communities are part of the solution. Each of the committees is supported by facilitation, planning, and technical (including scientists) staff to help them understand issues and develop potential solutions. The Committees and support staff are funded through general regional taxes.

Figure 3.3. **The stakeholder engagement staircase**

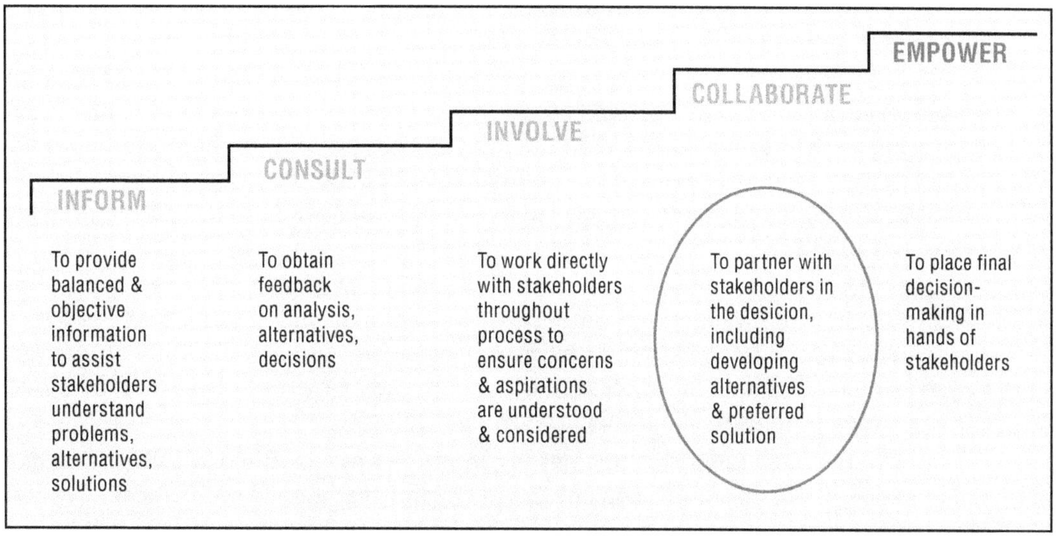

Source: Case study submitted in full by Nic Newman, Environment Canterbury, New Zealand.

Once committees have developed their Zone Implementation Programmes and submitted them to the Canterbury Regional Council and district and city councils, they are given effect to through any one, or a combination of, the following three pathways:

- Regional plans under the New Zealand Resource Management Act (1991) to give effect to statutory recommendations e.g. catchment load limits for water quality;
- Agency work programmes to give effect to non-statutory recommendations e.g. stream restoration, or irrigation efficiency programmes; and
- Infrastructure work programmes and matching central and regional government funding to give effect to recommendations related to infrastructure and restoration.

As an example, the Upper Waitaki Zone Committee (established in 2010), with the help of the local community and technical support (with expertise in economics, cultural values, social science, modelling, water quality and ecology) developed a Zone Implementation Programme comprising of: *i)* desired community water quality outcomes; *ii)* recommendations for water quality limits based on maintaining the trophic state of Lake Benmore; *iii)* catchment nutrient loads for all activities; *iv)* the method of allocating the nutrient loads; *v)* methods to incentivise biodiversity protection (e.g. an easier resource consent pathway for development that is accompanied by biodiversity protection); *vi)* non-statutory actions such as an education campaign for visitors; *vii)* a rehabilitation programme for degraded water bodies; and *viii)* an integrated monitoring framework for the Zone to track progress and to share data.

In terms of nutrient loading, the technical assessment indicated that any further development (that resulted in an increase in nitrogen loss) beyond what was consented in the Ahuriri Arm of Lake Benmore, would risk the lake moving from an oligotrophic state to a mesotrophic state. The community considered this risk to be unacceptable. In accordance with this decision, the decision made to "grandparent" modelled nitrogen losses to individual land owners based on current land-use at "good management practice", and then use a "modified equal" method of nutrient allocation where there was room for intensification. This solution sought to make the best use of the potential for further development by allowing intensification on the most productive land in an equitable way without comprising water quality outcomes.

The Collaborative Governance Model not only resolved how to set water quality (and quantity) limits and other actions to deliver on the CWMS targets, but also facilitated delivering on the National Policy Statement for Freshwater Management. Recommendations made to the Canterbury Regional Council by Zone Committees for specific planning provisions have been implemented through plan changes to the Canterbury Land and Water Regional Plan (2015).

Six of the ten Zone Committees have now been through a rigorous process and reached consensus on water quality and water quantity limits for their zones. There is widespread agreement that there is no going back and collaboration is the only way forward; the paradigm has changed. The success of the collaborative approach is now spilling over in to other sectors, such as public transport governance.

One of the most tangible outcomes is community ownership of solutions. For example, in one zone, the community (via the Zone Committee) agreed to increase local taxes to support the maintenance of a key water infrastructure asset now that the wider benefits provided by this asset are more clearly understood. In another zone, the community have taken ownership of lagoon augmentation as part of the local solution and they (not the council) are working together to figure out how to design and pay for it. When people are part of developing solutions, they are invested in seeing them come to fruition. An outcome most frequently quoted by participants is the personal development of their own skills and knowledge, and appreciation for alternative views. Additionally, the relationships and social capital built has enabled progress to be made on issues that have been stuck for years that are outside of the CWMS, for example dealing with gravel aggradation in a river. By bringing people together to solve problems, the sum is greater than the total of the parts.

Challenges associated with a collaborative governance process include speed of implementation and delivery; inherently, collaborative processes move at the "speed of trust". There is variable capacity of community members to understand and assimilate information that includes complex biophysical, cultural, social and economic data. Also, community members have to dedicate considerable time to the collaborative process (time commitment can vary from a minimum of at least half a day a month to fortnightly meetings and workshops, with pre-reading). Added to this, committee members are often exposing themselves to their community by fronting difficult conversations and solutions. Finally, if safeguards are not put in place to ensure all stakeholders have equal representation, disproportionate capture of vested interests in collaborative groups can reduce the potential to achieve ambitious water quality limits in a timely manner.

Based on the experience of the CWMS, requisites for a successful collaborative governance approach include:

- *Objective and clarity of the process.* The CWMS set out the principles, targets, and methodology "up-front" and removed any doubt over scope, process, and what was trying to be achieved.

- *Commitment and clarity from governors on the lines of decision making*. The Canterbury Regional Council delegated significant power to the zone committees by agreeing to endorse all of the committees' recommendations where these are the consensus of the committee and have been developed with strong stakeholder and community engagement.
- *Absolute transparency with information and process*, including having difficult conversations in sessions that are open to the public and making all technical information freely available. Traceability is important; the wider community needs to be able to know when, where, why and how, certain decisions were made. It is critical to get right, and be clear about, the scale of operation - hydrological, social, and administrative.
- *Resourcing needs to match the level of ambition for stakeholder engagement and responsibility*. The most substantial expenditure is the support staff. Facilitators need to be able to deal with ambiguity, to think and work across disciplines, and be committed to developing resolutions but not transposing their own ideas (i.e. to be "knowledge brokers"). Technical support staff who provide science, hydrology, planning, biodiversity, cultural and infrastructure advice need to be able to communicate at various levels and the facilitators need to be prepared to "hold a space" for stakeholders who may not be well resourced or articulate.

Notes

1. Refer to Chapter 1 for an explanation of the different characteristics of point versus diffuse source pollution.
2. Fertiliser taxes can cause an additional burden on horticulture production while making livestock production more profitable. They may also provide unintended incentives to increase livestock levels, leading to greater manure production through more intensive protein feeding, larger acreages devoted to nitrogen-fixing plants and reorganisation of crops in favour of those with less nitrogen consumption, but not necessarily less nitrogen surplus (OECD, 2010).
3. Summary of case study provided in full by Jihyung Park and Sang-Cheol Park, Ministry of Environment, Korea.
4. Summary of case study provided in full by Dr. Alec Mackay, AgResearch, New Zealand.
5. The One Plan is the new regional plan to guide the management of natural resources in the Manawatu-Wanganui region of New Zealand. It is called the One Plan because it weaves together resource management plans (air, land, water quantity, water quality, biodiversity, coastal, natural hazards) and the Regional Policy Statement into one easy-to-use document. The Plan became operative on 19 December 2014. www.horizons.govt.nz/about-us/one-plan/.
6. OVERSEER® is a software application that supports farmers and growers to make informed strategic management decisions about their nutrient use on-farm to improve performance and reduce losses to the environment. New Zealand farmers will increasingly use the model to develop nutrient management plans and budgets, as required by regional councils. The model does not include economic analysis, so outputs needs to be combined with other economic models to assess impacts of options on the farm business. http://overseer.org.nz/.
7. For details, www.hbrc.govt.nz/hawkes-bay/projects/tukituki/plan-change-6/.
8. Water quality trading is a market-based mechanism that allows sources with high pollution control costs to purchase credits, or pollution discharge reductions, from sources with lower pollution control costs.
9. Summary of case study provided in full by Sara Walker, World Resources Institute, DC, United States.
10. Case study sourced from: OECD (2017), *OECD Environmental Performance Reviews: New Zealand 2017*, OECD Publishing, Paris, http://dx.doi.org/10.1787/9789264268203-en.
11. Summary of case study submitted in full by Jim Gebhardt, Environmental Protection Agency, United States.
12. Summary of case study provided in full by Nick Haigh, Defra, United Kingdom.
13. Summary of case study submitted in full by Nic Newman, Environment Canterbury, New Zealand

References

Chesapeake Bay Foundation (2014), The economic benefits of cleaning up the Chesapeake: A valuation of the natural benefits gained by implementing the Chesapeake clean water blueprint.

Daigneault, A., S. Greenhalgh and O. Samarasinghe (2017), Equitably slicing the pie: Water policy and allocation, *Ecological Economics*, Vol. 131, pp. 449–459.

Defra (2015), Costs and benefits of the Thames Tideway Tunnel, 2015 update https://www.gov.uk/government/uploads/system/uploads/attachment_data/file/471845/thames-tideway-tunnel-costs-benefits-2015.pdf [accessed 10/03/2016].

Duhon, M., H. McDonald and S. Kerr (2015), "Nitrogen trading in Lake Taupō: An analysis and evaluation of an innovative water management policy, *Motu Working Paper, Vol.15-07,* Motu Economic and Public Policy Research, Wellington.

Eftec (2015), Update of the Economic Valuation of Thames Tideway Tunnel Environmental Benefits, Final Report for Defra, August 2015.

Farrow, R.S. Schultz, M.T., Celikkol, P. and G.L. Van Houtven (2005), Pollution trading in water quality limited areas: use of benefits assessment and cost-effective trading ratios, *Land Economics*, Vol. 81, No. 2, pp. 191-205.

Fishmana, Y., Beckerb, N. and M. Shechterc (2012), The Polluter Pays Principle as a policy tool in an externality model for nitrogen fertilizer pollution, *Water Policy*, Vol. 14, pp. 470–489.

Grolleau, G. and L.M.J. McCann (2012), "Designing watershed programs to pay farmers for water quality services: Case studies of Munich and New York City", *Ecological Economics*, Vol. 76(2012), pp. 87-94.

Hanley, N., J.F. Shogren and E. White (2013), The Economics of Water Pollution, In: Introduction to *Environmental Economics*, second Edition, Oxford University Press, Oxford, UK.

Hoffmann, S.J. and J. Boyd (2006), Environmental Fees: Can incentives help solve the Chesapeake's nutrient pollution problems, Resources for the Future, Washington, D.C.

Hoffmann, S., J. Boyd and E. McCormick (2006), Taxing nutrient loads, *Journal of Soil and Water Conservation*, Vol. 61, No. 5, pp. 142A-147A.

Horizons Regional Council (2014), One Plan: The Consolidated Regional Policy Statement, Regional Plan and Regional Coastal Plan for the Manawatu-Wanganui Region. https://www.horizons.govt.nz/publications-feedback/one-plan [accessed 04/04/2016].

Jones, C. (2010), How Nutrient Trading Could Help Restore the Chesapeake Bay, WRI Working Paper. World Resources Institute, Washington, DC.

Kim, E.J. et al. (2016), The effect of total maximum daily loads (TMDL) program on water quality improvement in the Geum River basin, Korea, *Desalination and Water Treatment*, Vol. 57(2), pp. 597-609.

Köhler, J. (2001), *Detergent phosphates and detergent ecotaxes: a policy assessment*, University of Cambridge, UK.

Lynn, I.H. et al. (2009), *Land use capability survey handbook – a New Zealand handbook for the classification of land* 3rd ed., pp. 163. AgResearch, Hamilton; Land case Research, Lincoln; and GNS Science, Lower Hutt.

Mackay, A.D. (2009). *Evidence IN THE MATTER of hearings on submissions concerning the Proposed One Plan notified by the Manawatu-Wanganui Regional Council.* http://old.horizons.govt.nz/assets/1plan_eoh-report/Dr-Alec-Donald-Mackay-End-of-Hearing-Technical-Report.pdf [accessed 04/04/2016] .

Montginoul, M., S. Loubier, B. Barraque and A. Agenais (2015), Water pricing in Frabce: Toward more incentives to conserve water, in: Dinar, A., V. Pochat and J. Albiac-Murillo (Eds.) Water Pricing Experiences and Innovations, Springer International Publishing, Switzerland.

Muller, N.Z. and R. Mendelsohn (2009), Efficient Pollution Regulation: Getting the Prices Right, *American Economic Review*, Vol. 99, No. 5, pp. 1714–1739.

National Audit Office (2010), Tackling diffuse water pollution in England, Report by the Comptroller and Auditor General, July 2010;

National Institute of Environmental Research (NIER) (2014), Technical Guidelines for TMDL Management, Korea.

OECD (2017), *OECD Environmental Performance Reviews: New Zealand 2017*, OECD Publishing, Paris. http://dx.doi.org/10.1787/9789264268203-en.

OECD (2015a), *Innovation, Agricultural Productivity and Sustainability in Canada*, OECD Food and Agricultural Reviews, OECD Publishing, Paris. http://dx.doi.org/10.1787/eco_surveys-nzl-2015-en

OECD (2015b), *OECD Economic Surveys: New Zealand 2015*, OECD Publishing, Paris. http://dx.doi.org/10.1787/eco_surveys-nzl-2015-en.

OECD (2015c), "The Lake Taupo Nitrogen Market in New Zealand: Lessons in environmental policy reform", *OECD Environment Policy Papers*, No. 4, OECD Publishing, Paris, http://dx.doi.org/10.1787/5jrtg1l3p9mr-en.

OECD (2015d), *Water and Cities: Ensuring Sustainable Futures*, OECD Studies on Water, OECD Publishing, Paris, http://dx.doi.org/10.1787/9789264230149-en.

OECD (2014), Green Growth Indicators 2014, OECD Green Growth Studies, OECD Publishing, Paris, http://dx.doi.org/10.1787/9789264202030-en.

OECD (2013a), *Water Security for Better Lives*, OECD Studies on Water, OECD Publishing, Paris, http://dx.doi.org/10.1787/9789264202405-en.

OECD (2013b), *Providing Agri-environmental Public Goods through Collective Action*, OECD Publishing, Paris, http://dx.doi.org/10.1787/9789264197213-en.

OECD (2012), *Water Quality and Agriculture: Meeting the Policy Challenge*, OECD Studies on Water, OECD Publishing, Paris, http://dx.doi.org/10.1787/9789264168060-en.

OECD (2010), *Taxation, innovation and the environment*, OECD Publishing, Paris, http://dx.doi.org/10.1787/9789264087637-en.

OECD (2004), "Reducing water pollution and improving natural resource management", *in: Sustainable Development in OECD Countries: Getting Policies Right*, OECD Publishing, Paris, http://dx.doi.org/10.1787/9789264016958-7-en.

Ofwat (2011), From catchment to customer: Can upstream catchment management deliver a better deal for water customers and the environment?, September 2011, Ofwat, UK.

Parris, K. (2012), Impacts of agriculture on water pollution in OECD countries: Recent trends and future prospects, *In: Water Quality Management: Present Situations, Challenges and Future Perspectives*, Eds. A.K. Biswas, C. Tortajada and R. Izquierdo, Routledge, Oxfordshire, UK.

Rejwan, A. and Y. Yaacoby (2015), Israel: Innovations overcoming water scarcity, *OECD Observer Business Brief* No. 302, OECD, Paris and Watech, Mekorot, www.oecdobserver.org/news/fullstory.php/aid/4819/Israel:_Innovations_overcoming_water_scarcity.html#sthash.KWiFgV3I.dpuf [accessed 05/04/2016].

Shortle, J. S. and R.D. Horan (2001), The economics of nonpoint pollution control, *Journal of Economic Surveys*, Vol. 15 (3), pp. 255 – 289.

Shortle, J.S. and R.D. Horan (2013), Policy Instruments for Water Quality Protection, *Annual Review of Resource Economics*, Vol. 5, pp. 111-138.

Shortle, J.S. et al. (2012), Reforming agricultural nonpoint pollution policy in an increasingly budget-constrained environment, *Environmental Science and Technology*, Vol. 46, pp 1316-1325.

Thames Tideway Tunnel (2015), https://www.tideway.london/the-tunnel/ [accessed 01/09/2015].

UK Environment Agency (2015), Impact Assessment; Update to the river basin management plans for England's water environment, October 2015;

US EPA (2011), Drinking Water Infrastructure Needs Survey and Assessment, Fifth Report to Congress, www.epa.gov/tribaldrinkingwater/drinking-water-infrastructure-needs-survey-and-assessment-fifth-report-congress.

Withana, S. et al. (2014), *Environmental tax reform in Europe: Opportunities for the future*, A report by the Institute for European Environmental Policy (IEEP) for the Netherlands Ministry of Infrastructure and the Environment.

Diffuse Pollution, Degraded Waters
Emerging Policy Solutions
© OECD 2017

Chapter 4

A policy framework for diffuse source water pollution management

This final chapter presents a policy framework for diffuse source water pollution management and concludes with recommendations for central government. The framework and recommendations are based on the outcomes from the OECD Workshop on Innovative Policy Responses for Water Quality Management, and draws upon the policy analysis of case studies throughout the previous chapters of the report.

Key messages

A policy framework can assist policy makers and stakeholders through the myriad of decisions required to establish new or alter existing water quality management regimes. A policy framework for diffuse water pollution management operates on three levels: *i)* political ambition; *ii)* policy principles; and *iii)* policy instruments. There is also a specific role for central government to help facilitate and expedite reductions in diffuse water pollution and improvements in water quality.

The first level of the framework – political ambition – stresses the importance of knowing and targeting diffuse water pollution risks. Completely eliminating water pollution risks is often technically impossible and not cost-effective; risks must be prioritised and policy responses targeted based on the acceptable level of risk for society (economic, social and environmental), and the cost of amelioration. A lack of full scientific certainty should not be used as a reason for postponing measures to prevent environmental degradation. Connecting with higher level policy issues can assist in triggering political action (e.g. health, food security, economics, energy production and tourism). New knowledge and tools are available to assist in reducing uncertainties for the development of policy responses.

The second level of the framework – policy principles – outlines a hierarchy of OECD principles for water quality management: The Principles of Pollution Prevention; Treatment at Source; Polluter Pays and Beneficiary Pays. Equity should be considered with regards to who the costs and benefits of policy reform fall upon and the needs of future generations. Policy coherence is required to ensure initiatives taken by different policy sectors (e.g. agriculture, urban planning, and climate) do not have negative impacts on water quality and freshwater ecosystems, or increase the cost of water quality management. A number of OECD Principles on Water Governance are also applicable with regards to: the management scale of diffuse pollution; data and information; implementation and enforcement of policies; and stakeholder engagement.

The third and final level of the framework – policy instruments – recognises that it is not economical to observe individual diffuse water pollution sources directly. Policy makers seeking to achieve environmental goals, while minimising transactions costs and the direct cost of diffuse pollution control must choose from three alternative management options with which regulations, economic instruments and/or voluntary mechanisms will apply to: *i)* managing land use practices as proxies; *ii)* rewarding or penalising polluters collectively for their jointly determined impacts on ambient pollution levels at particular receptors; or *iii)* managing estimated diffuse emissions via computer modelling. The third option offers an opportunity to design policy instruments directly proportional to the amount of estimated pollution generated or reduced from individual properties as part of a wider catchment.

Greater use of economic instruments is required to effectively manage diffuse water pollution. In particular, better utilising economic instruments (e.g. pollution charges, taxes or water quality trading) can create incentives to reduce pollution, and increase the cost effectiveness of and innovation in pollution control strategies.

Central government has a critical role to play in the transition to more effective management of diffuse water pollution. Recommendations include: *i)* providing overarching national policy guidance and minimum standards; *ii)* creating the institutional framework setting the distribution of responsibilities across levels of government; *iii)* stakeholder engagement on approaches to manage perceived and actual risks, and a commitment to reach solutions in partnership; *iv)* signalling policy changes and highlighting options for implementation; and *v)* stimulating the diffusion of innovative technical and policy approaches that minimise the cost of water quality management (including seed funding, space for experimentation and making pollution costly). Lastly, monitoring, enforcement and evaluation of policy implementation, ongoing stakeholder engagement, and reassessment of the risks, are necessary in order to adapt to future changes in climate, economic growth, population dynamics and advances in science and technology.

A policy framework for diffuse source water pollution management

As discussed in the previous chapter, policy approaches used to date for the management of water pollution have largely focussed on point source pollution control with large investments in wastewater treatment, and a reliance on voluntary participation and compliance measures for diffuse sources of pollution (Shortle, 2017). However, water quality remains an issue of concern in OECD countries. Governments have struggled to implement policies that successfully reduce pollution from diffuse sources of pollution. This limited success reflects the inherent complexity of diffuse pollution, and political resistance to regulation and application of the polluter pays principle.

While there is no silver bullet or one-stop-shop for effective diffuse water pollution management, this chapter presents a policy framework (outlined in Table 4.1) that provides a structure to support policy makers to make more robust and defensible decisions. Recommendations for central government to help facilitate and expedite reductions in diffuse water pollution and improvements in water quality are also provided.

Table 4.1. **A policy framework to manage diffuse water pollution**

Level	Description
Political ambition	*Know the risks* Identify pollutants, sources, pathways, timing and sensitivity of the receiving environment. Assess the diffuse water pollution risks (environmental, economic and social) taking into account time lags, historical pollution and planned land use change. *Target the risks* Limiting diffuse pollution comes at a cost. Set the appropriate level of risk and ambition and determine priorities informed by thorough assessments, robust knowledge and stakeholder engagement.
Policy principles	Hierarchy of principles for action: • Principle of Pollution Prevention • Principle of Treatment at Source • Polluter Pays Principle • Beneficiary Pays Principle Consider Equity with regards to who the costs and benefits of policy reform fall upon and the needs of future generations. Encourage Policy coherence across sectors that affect diffuse pollution. Ensure good water governance, with reference to the OECD Principles on Water Governance, in particular: geographical scale; data and information; implementation and enforcement; and stakeholder engagement and outcome-oriented contributions to policy design.
Policy instruments	*Manage the risks* Because it is not economical to observe diffuse water pollution directly, the choice and design of policy instruments should build upon one of three alternative management options: • Manage land use practices and inputs as proxies • Reward or penalise polluters collectively • Manage estimated diffuse emissions via modelling. Develop policy responses proportional to the magnitude of the risk. Target adoption of low cost strategies that achieve a high benefit return. Include local differences in the land resource (e.g. their ability to filter and retain water and pollutants) as an integral part of policy development. Consider economic instruments (e.g. pollution charges, product charges, and water quality trading), in combination with regulatory and voluntary mechanisms.

Political ambition

Know diffuse pollution risks

A risk-based approach to water quality utilises the OECD water security risk framework - "Know the risks", "Target the risks" and "Manage the risks" (OECD, 2013) (Figure 4.1).

Figure 4.1 **A Risk-Based Approach to water quality management**

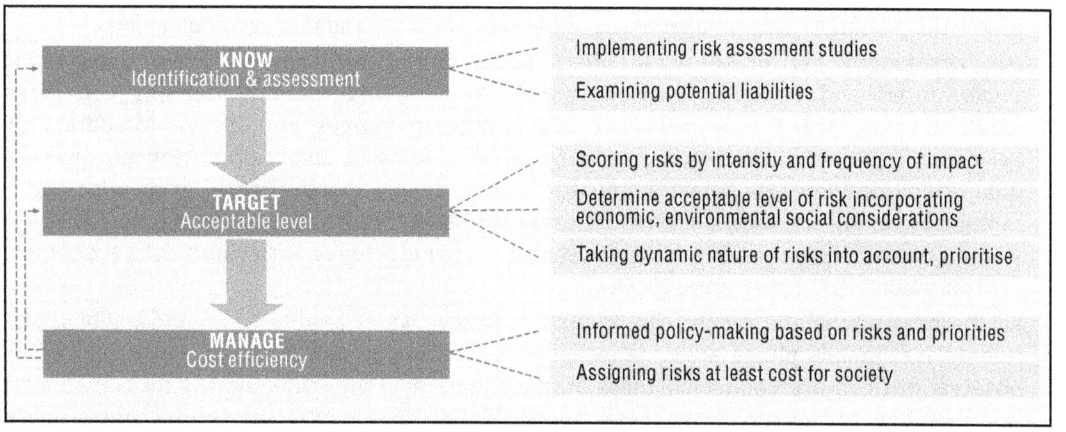

Source: Adapted from OECD (2013).

The first step in a risk-based approach to water quality is to identify pollutants, their origin, timing and pathways, and their risks to water quality, including their likelihood and impact. This will involve understanding the relative contribution of both point and diffuse sources of pollution. Knowledge and information are necessary to understand the causes and impacts, both short- and long-term, and to assess the hazards, exposure, and vulnerability of people and assets. Time lags in the system of diffuse pollution from current and historic land use add complexity to the management of water quality.

The identification and management of diffuse water pollution would benefit from improved scientific knowledge and understanding; greater understanding of the scientific and economic relationships will allow policies to be better targeted, informed and refined. New knowledge and tools are available to assist in this process and should be harnessed (Box 4.1). Stakeholder engagement is an important component of knowing the risks of diffuse water pollution, and understanding the risk perceptions of stakeholders.

Box 4.1. **Make the best of new knowledge to manage diffuse water pollution**

Diffuse water pollution is considered difficult to manage because of the challenges to regulate large numbers of small sources and because the science associated with assessing the impact of each diffuse source is complex (Anastasiadis et al., 2013). And yet, a science-based approach is essential to formulate a multidisciplinary approach to the complex issue of water quality; to manage water quality and quantity in unison, their shocks and tipping points, and their spill-overs to other locations, media (i.e. water, air, land) and sectors (Grey et al., 2013). New knowledge and tools are available to assist in this process.

Advances in computer modelling enables estimation of diffuse pollution based on farm practices (such as crop rotations, stocking ratios, tillage practices, fertiliser, pesticide and irrigation applications), and the hydrological, soil and geographical conditions that effect the transport of pollutants to surface and groundwater bodies (Fishmana et al., 2012). With modelling, diffuse pollution can be projected from individuals to the scale of the catchment, and thereby offers an opportunity for diffuse pollution to be managed as a "point source", with individual land owners held responsible for their actions, and individual land parcels managed as part of a wider catchment to achieve water quality objectives.

Box 4.1. **Make the best of new knowledge to manage diffuse water pollution** *(cont.)*

More monitoring often identifies more problems. Computer modelling can offer a way forward by identifying pollution hotspots and pollution source priorities. By merging physical water quality models with economic models and sensitivity analysis, the efficiency and effectiveness of "what-if" scenarios of various policy and infrastructure options can be tested without recourse to expensive testing in reality (Anastasiadis et al., 2013). Such a decision-support tool can save time and resources, assess the potential risks to stakeholders and prioritise policy and management actions.

An increasingly networked world offers opportunities for capturing new data, reducing uncertainties and engaging with the public. Remote and real-time sensing can generate new knowledge of the state of water quality, pollution sources, and options to address them. For example, Korea's water agency, K-Water is responding to uncertainty in water quality and quantity by developing a Smart Water Grid that combines existing water grids with real time monitoring to ensure adequate quantity and consistent water quality, and to detect leakages in water systems, thereby maximising water and energy efficiency with significant economic and environmental benefits (Brears, 2016).

Earth observations and drones can be used to assess water quality in remote regions. Citizen science (i.e. mobile phone apps, online pollution reporting and pollution hotlines) and earth observations can overcome challenges of inadequate data and data sharing for transboundary management. For example, the Creek Watch iPhone App enables the United States public to capture data on the quality and quantity of any water body at any point and time (IBM, 2012). The App enables new sources of data to be collected at little added cost, from which new insights can be derived and management decisions prioritised. Some challenges associated with using citizen science strategically include: integration and coherence of data gathered from various citizen and other sources; that data gathered is also made available to citizens, scientists, regulators, and polluters in accessible and understandable ways; and that citizen science efforts and online platforms for accessing data are sustained beyond typical three-year project funding cycles.

All of these new sources of data have the ability to reduce monitoring, compliance and enforcement costs. The new digital environment also offers opportunities for more collaborative and participatory relationships that allow relevant stakeholders to actively shape political priorities, collaborate in the design of public services, and participate in their delivery to provide more coherent and integrated solutions to complex challenges (OECD, 2016a). The benefits of providing free and publically accessible data on water quality are recognised by several OECD countries. For example, the United States Water-Quality Watch website displays real-time water quality data collected remotely by sensors installed in rivers, lakes, and other water bodies, and the Rivers and Streams Water Quality Website presents interactive annual graphical summaries of streamflow information, and nutrient and sediment concentrations and loads. Users can compare recent and long-term water quality conditions, download data, evaluate nutrient loading to coastal areas, and more (USGS, 2015). New Zealand has established a similar national public database (lawa.org.nz) to improve utilisation of reliable real-time and historical water quality data for a range of users to inform business, recreational and environmental decisions. Sharing data frees up significant overheads in delivering routine data requests, avoids double-monitoring, and redirects effort into additional monitoring or policy work.

Sources: Anastasiadis et al. (2013); Fishmana et al. (2012); Vörösmarty et al. (2010); Brears (2016); IBM (2012); OECD (2016); Grey et al. (2013).

Target diffuse pollution risks

Completely eliminating water pollution risks is often technically impossible and not cost-effective (OECD, 2013). The second step of the risk-based approach is prioritising and targeting selected water quality risks. This involves determining the acceptable level of water risk for society, depending upon the balance between economic, social and environmental consequences, and the cost of amelioration. Action should be targeted on pollutants of particular significance at the scale and sensitivity of the catchment, basin, or aquifer, on the basis of characteristics such as toxicity, persistence and bio-accumulation (see typology of water pollution, Figure 1.2, Chapter 1). Threats of serious or irreversible environmental damage should be prioritised.

Risk assessments coupled with cost-benefit analyses and computer modelling can help in ranking priorities, identifying the high risk pollution hotspots, and the policy responses most likely to achieve the greatest societal benefit under a range of potential future scenarios. Such tools can be used to support decision-making, but in the end, decisions have political influence which may be informed by the following criteria:

- Stakeholder engagement to help determine the level of acceptable risk to society
- Economic impacts, including cost of amelioration and burden sharing
- Human health and social impacts
- Impact on freshwater biodiversity, ecosystems and ecosystem services
- Geographical extent of impacts
- Longevity and irreversibility of impacts.

A lack of full scientific certainty should not be used as a reason for postponing measures to prevent environmental degradation. Connecting with higher level policy issues can assist in triggering political action (Box 4.2).

Whether land is used for intensive agriculture, forestry, conservation, roads, city living or other purposes, the interaction between the land and other production inputs inevitably changes the physical and chemical characteristics of water. Including local differences in the land resource itself (e.g. its ability to filter and retain water and pollutants) should be an integral part of diffuse water pollution risk assessment and policy development.

The third and last step of the OECD risk-based approach involves managing the risks, which is discussed in level of "policy instruments" of the Framework.

Box 4.2. **Connect with higher level policy issues to trigger political action on water pollution**

In order to trigger political action to set and enforce regulations, and to raise stakeholder awareness, marketing the importance of water quality objectives beyond environmental protection can be a strategic way forward. Successful initiatives often connect water quality issues with issues of higher political value, or are readily associated with obvious benefits, such as health, ecosystem services, economics and food security. A compelling case for action can be made to central and local governments, and to the public and stakeholders, by highlighting the co-benefits valued by policy leaders and public opinion. The table below lists a range of co-benefits to governments for improved water quality management.

Box 4.2. Connect with higher level policy issues to trigger political action on water pollution *(cont.)*

Identifying touchstones can make tangible connections with water quality improvements and support compelling stories. For example, restoration of otters, salmon and other fish in England's rivers has triggered a strong political coalition to improve water quality (Defra, 2011). Making the economic case, computing the cost of inaction, and strengthening valuations of diffuse water pollution in environmental impact assessments can support proposals for action. Information campaigns can rally public support.

Examples of co-benefits to governments for improved water quality management

Co-benefits of improving and protecting water quality	
• Secure water resources (increase water of useable quality). • Adapt to a changing climate. • Reduce flood risk (e.g. catchment protection management, permeable pavements, swales, wetlands). • Reduce health costs (e.g. cancers associated with nitrates). • Improve biodiversity, ecosystem health and the value of ecosystem services. • Improve long term sustainable agricultural, aquaculture and industrial productivity. • Energy production and utilisation of finite minerals and nutrients through wastewater reuse. • Increase cultural and social relations and trust in government.	• Reduce economic losses associated with sick days. • Sustain and increase food security. • Boost and protect tourism. • Improve marketing image/reputation for exports. • Reduce energy consumption (pumping stormwater, water treatment). • Reduce water treatment costs (reduce the need for upgrades or additional plants). • Improve amenity of waterfront properties and public spaces. • Mitigate climate change (e.g. forested catchments, wetlands and build-up of soil organic matter). • Increase recreational use of water bodies. • Increase value of water resources (economic, cultural, spiritual, environmental) and human wellbeing.

Sources: Authors own; Defra (2011).

Policy principles to guide decision-making

The following set of OECD principles can usefully guide the development of policy for the management diffuse pollution sources. They are captured by the *Recommendation of the OECD Council on Water* (OECD, 2016b):

Hierarchy of principles for action

The Principle of Pollution Prevention reflects that prevention of diffuse pollution is often more cost effective than treatment/restoration options. This means preventing pollutants from reaching water bodies by means such as recovery and re-use of wastewater, product substitution, modification of industrial processes, best land management practices and retirement of land.

The Principle of Treatment at Source considers that pollution control measures should be applied as close to the source as possible. In effect, the later the stage of control, the less effective it is likely to be due to wider dispersion of the contaminants. Particularly strict measures of control should be enforced for certain categories of hazardous pollutants with a view to preventing their dispersion into the environment. This applies especially to toxic substances which are persistent in the environment and/or subject to bioaccumulation in living organisms and concentration through the food chain (e.g. heavy metals, DDT). Management measures should aim to prevent uncontrolled pollution transfers to other water resources, or to soil or atmospheric systems.

The Polluter Pays Principle creates conditions to make pollution a costly activity and to either influence behaviour to reduce pollution, or generate revenues to alleviate pollution and compensate for social costs (OECD 2012a). Examples include pollution charges, taxes on inputs (such as fertilisers and pesticides) and sewer user charges. The polluter pays principle should not be accompanied by conflicting subsidies, tax advantages or other measures that encourage polluters to pollute, or assist polluters in bearing the costs of pollution, thereby creating distortions in the market (OECD, 1972; 1974). While there is a case for a public subsidy to address the accumulated damage caused by historical pollution (particularly when the polluters are no longer around to pay), the polluter pays principle should be the first line of defence in securing water quality and incentivising behaviour change (e.g. through water pollution charges and water quality trading).

There are several challenges that result in the polluter pays principle not frequently being applied in the control of diffuse pollution (it is more commonly used with the control of point source pollution) (Table 2.7, Chapter 2). They include difficulties with identifying and targeting polluters, determining reliable estimates of pollution costs, poor enforcement of existing regulations, and strong political opposition. Possible ways to overcome these barriers are captured in Table 4.2 below.

Table 4.2. **The Polluter Pays Principle for diffuse water pollution: Barriers and solutions**

Barriers	Solutions
Difficulties with identifying and targeting polluters	• Computer modelling as a cost-effective alternative to directly observing individual diffuse pollution emissions • Taxes on inputs (e.g. fertilisers, pesticides, cleaning products) or land use (e.g. paved urban surfaces, livestock numbers, intensive land use) • Collective accountability at catchment level
Difficulties with determining reliable estimates of pollution costs	• Economic modelling and scientific monitoring to inform costs and justify action (new data sources are available, see Box 4.1) • Market mechanisms to reveal pollution costs and differentiated abilities to cope with them
Poor enforcement of existing regulations	• Computer modelling as a cost-effective alternative to directly observing individual diffuse pollution emissions • Taxes on inputs (e.g. fertilisers, pesticides, cleaning products) or land use (e.g. paved urban surfaces, livestock numbers, intensive land use) • Collective accountability at catchment level • Increased financial and technical support for local authorities to enforce regulations
Strong political opposition	• Economic modelling and scientific monitoring to inform costs and justify action (new data sources are available, see Box 4.1) • Stakeholder engagement • Collective accountability at catchment level • Connecting with higher-level policy priorities

The Beneficiary Pays Principle allows sharing of the financial burden of water quality management. It takes account of the high opportunity cost related to using public funds for the provision of private goods that users can afford. A requisite is that private benefits attached to water resources management are inventoried and valued, beneficiaries are identified, and mechanisms are set to harness them (OECD 2012a). For example, wastewater treatment plants help to protect water quality in rivers and lakes, and green infrastructure, such as wetlands and forested catchments, provide water filtration ecosystem services. Beneficiaries include city residents provided with quality drinking water; reduced water treatment costs for utilities and health systems, and downstream industrial and agricultural users; improved business for fisheries and tourism operators; and benefits for recreational users, waterfront property owners, the environment, and society at large. Compliance with baseline regulations must be achieved before a payment for ecosystem service scheme is implemented. This is required to ensure additionality and to prevent polluters being rewarded.

Additional principles for policy design

Equity should be considered with regards to who the costs and benefits of policy reform fall upon, and the needs of future generations. Disproportionate costs to users, while important, should not be overstated. Where high levels of taxes have been applied to chemical inputs to comply with the polluter pays principle, often coupled with a mix of other policy measures, they have usually led to reductions in input use without loss of farm production or income (OECD, 2012b). Due consideration of the equity principle for water quality management financing should also be given for public subsidies (OECD, 2009). Equity and fairness in burden sharing do not preclude efficiency.

Policy coherence is required to ensure initiatives taken by different policy sectors do not have negative impacts on water quality and freshwater ecosystems, or increase the cost of water quality management. Multiple policy sectors affect diffuse water pollution and its management, for example, urban development, agriculture, climate, natural resources, forestry, energy, conservation and human health. Policy coherence would entail:

- The removal of subsidies that encourage land use change or intensification that can result in diffuse water pollution.
- Looking for win-win solutions such as NOx reductions to improve air and water quality and reduce greenhouse gas emissions simultaneously.
- Integrating water pollution control (both point and diffuse source) with air pollution control, land use management, and water quantity management.

In "knowing", "targeting", and "managing" water quality risks, identifying trade-offs and impacts is critical. In particular, water quality and water quantity should be managed in unison as the two are interrelated and interdependent. For example, poor water quality reduces the quantity of useable water and therefore exacerbates the problem of water scarcity. Water scarcity reduces the capacity for dilution of point source pollution. High rainfall events and flooding cause diffuse pollution from land runoff (agricultural and urban) and combined sewer overflows into rivers.

In addition, there may be trade-offs and co-benefits between water quantity and quality management, and other important sectoral policies, such as land, energy, biodiversity, urban planning, health care, waste, construction, transport, and climate change. Increasing desalination to improve water security requires large amounts of energy and produces highly concentrated brine. Intensifying land use for food security requires greater inputs of water, energy and nutrients, and contributes to water pollution and climate change.

The potential synergies and complementarities among the sectors should be used to guide formulation of effective options to maximise gain, optimise co-benefits, and avoid negative impacts. Examples of the potential trade-offs and co-benefits from water quality interventions are provided in Table 4.3. Similarly, there are benefits of factoring water quality into policies that affect water availability and use.

When considering new policy in other sectors that may have potential impact on water quality (e.g. agriculture, urban development, energy, climate, mining, etc.), it is important that their impacts on water quality, freshwater ecosystems, the economy and social welfare, and their underlying causes (e.g. market, information, institutional and enforcement failures, and perverse subsidies) are identified. Strengthening valuations of diffuse water pollution in environmental impact assessments can assist with the identification of trade-offs and co-benefits. The decision to commit to a new policy can be guided by a benefit-cost framework that measures whether the potential benefits of water quality protection, adjusted to account for risks, outweigh the potential costs. International experience and lessons learned from previous policy successes and failures should be applied. Evaluating the impact and effectiveness of new policy after implementation (ex ante) is equally important.

Table 4.3. **Examples of water quality policies and trade-offs and co-benefits to other sectors**

Water quality intervention	Potential trade-offs and co-benefits
Wastewater reuse to avoid pollution of rivers	Trade-offs: reduced environmental flow of rivers, additional energy requirements to process and/or transport wastewater and sludge from surplus regions to regions with a deficit. Co-benefits: utilisation of finite resources, such as phosphate, increased water security.
Higher drinking water quality standards to improve human health	Trade-offs: increased energy and chemicals consumption associated with increased water treatment, and increased carbon footprint. Co-benefits: reduced health costs.
Conversion to decentralised water and wastewater systems	Co-benefits: reduced energy consumption and carbon footprint from pumping water over large distances.
Restoration of wetlands	Co-benefits: reduced water treatment and energy consumption, increased biodiversity, carbon capture and storage, reduced flood risks.
Soil conservation to prevent erosion and sedimentation	Co-benefits: increased land use efficiency, biodiversity, food production, and water and fertiliser efficiency.

Principles on Water Governance

The OECD Principles on Water Governance (OECD, 2015a) can guide institutional arrangements for diffuse water pollution control. Four deserve particular attention:

- **Manage water at the appropriate scale(s) within integrated basin governance systems to reflect local conditions.** Because diffuse pollution is largely linked with hydrological processes, the catchment or basin scale is often the best scale for management. However, there may be atmospheric sources of pollution which do not conform to catchment or basin scales which should be considered. Cooperative transboundary water quality management may be required when water pollution affects another riparian country or state in a significant manner.
- **Produce, update, and share timely, consistent, comparable and policy-relevant water and water-related data and information,** and use it to guide, assess and improve water quality policy. Review data collection, use, sharing and dissemination to identify overlaps and synergies and track unnecessary data overload. Promote regular monitoring and evaluation of water quality policy. Develop reliable monitoring and reporting mechanisms to effectively guide decision-making and make policy adjustments when needed.
- **Ensure that sound water management regulatory frameworks are effectively implemented and enforced in pursuit of the public interest.** Develop a coherent legal and institutional framework that sets rules, standards and guidelines for achieving water quality policy outcomes, and encourages integrated long-term planning. Set clear, transparent and proportionate enforcement rules, procedures, incentives and tools (including penalties and rewards) to promote compliance and achieve regulatory objectives in a cost-effective way.
- **Promote stakeholder engagement for informed and outcome-oriented contributions to water quality policy design and implementation.** This is a key role for central and local government in the management of diffuse water pollution. The importance of stakeholder engagement is outlined in section 4.3.

Risk management and selection of policy instruments

Manage diffuse pollution risks

The third and last step of the OECD risk-based approach (Figure 4.1) involves managing the risks, and assigning risks to achieve the selected level of risk in the most equitable and cost-effective way.

It is not economical to observe individual diffuse water pollution sources directly. Policy makers seeking to achieve environmental goals, while minimising transactions costs and the direct cost of diffuse pollution control must choose from three alternative management options with which regulations, economic instruments and/or voluntary mechanisms will apply to:

- **Manage land use practices** (e.g. stormwater, nutrient management and erosion control practices) and inputs (e.g. fertilisers, irrigation) as proxies that can cause distribution of diffuse emissions. This is the most commonly used management approach for voluntary mechanisms to control diffuse water pollution. If applied to regulatory or economic policy instruments, it can limit land use practices and innovation, and can be less effective at reducing pollution in some instances (OECD, 2010).
- **Reward or penalise polluters collectively** for their jointly determined impacts on ambient pollution levels at particular receptors. This approach transfers the burden of asymmetric information and the difficulties of the measurement of ambient diffuse pollution and predictions under certain management scenarios from regulators to individual polluters.
- **Manage estimated diffuse emissions via modelling.** Computer modelling offers an opportunity for individual land parcels to be managed as part of a wider catchment to achieve water quality objectives. Policy measures to reduce diffuse pollution can be directly proportional to the amount of estimated pollution generated or reduced. It allows land managers to innovate farm and land management practices within a pollution limit without being restricted by the inputs and land use practices they use. However, the approach relies on a robust calibrated and validated model and reliable input data. This approach is discussed in application to economic instruments in the section below.

Policy responses should be proportional to the magnitude of the risk (OECD, 2013). When considering which particular policy instruments should be used to meet a given target for a water quality risk, an assessment should be made of how each instrument, or mix of instruments, is likely to contribute to the goals of water quality, economic efficiency and social equity. Risk assessments coupled with cost-benefit analyses and computer modelling can help in ranking priorities, identifying the high risk pollution hotspots, and the policy responses most likely to achieve the greatest societal benefit under a range of potential future scenarios. The risks, costs and benefits should be assigned according to the OECD policy principles outlined in the previous section. The adoption of low cost strategies that achieve a high cost-benefit return should be targeted.

Monitoring and evaluation of policy implementation and reassessment of the risks are necessary in order to adapt to changes; changes to the climate, population dynamics, economic growth, ageing infrastructure, evolving priorities and advances in science and technology make achieving and maintaining water quality a moving target. Criteria for assessing the viability and success of water quality policy reform may include: environmental effectiveness, economic efficiency, equity, administrative feasibility and cost, and acceptability.

Economic instruments as part of an effective policy mix

A policy mix to manage diffuse water pollution is required for effective and sustainable outcomes (economic, social and environmental). However, the present mix of regulatory and non-regulatory instruments in OECD countries limits the ability to address key pressures on water quality in the most cost-effective way. A number of examples of policy instruments to manage diffuse water pollution are presented in Table 4.4 and discussed in Chapter 3. In particular, economic policy instruments are under-utilised, although government interest in them is growing (Shortle, 2017). For sectors that increase aggregate amounts of water pollution, it is especially important that environmental externalities and opportunity costs be internalised where possible through the polluter pays principle. The addition of economic instruments (such as pollution taxes, charges, and water quality trading) would be one important step towards an effective policy mix.

Table 4.4. **Policy instruments to address diffuse water pollution and protect freshwater ecosystems**

Water-related risk	Regulatory	Economic	Voluntary or information-based
Water pollution	Water quality standards Mandatory best environmental practices and restrictions on inputs Pollution discharge permits Non-compliance penalties – non-renewal of resource permits or greater restriction on current permits Non-compliance fines	Pollution taxes (on inputs) Pollution charges (on outputs) Water quality trading Payment for ecosystem services	Information and awareness campaigns Farm advisory services for improved farming techniques (to minimise negative impacts on water quality) Contracts/bonds (e.g. land retirement contracts) Best environmental practices (or good management practices) Environmental labelling – products that meet certain environmental standards can be marketed and sold at a premium and/or subsidised.
Risk to the resilience of freshwater ecosystems	Minimum environmental flows (also for pollution dilution) Specification obligations relating to return flows and restrictions on point source discharges and irrigation in drought conditions	"Buy-backs" of water pollution allowances to ensure adequate water quality for ecosystem functioning	Information and awareness campaigns Voluntary surrender of pollution discharge allowances

Advances in nutrient pollution modelling provides an opportunity to utilise diffuse pollution charges, rather than taxing inputs as proxies (e.g. fertiliser use and livestock numbers), which can be less effective at reducing pollution[1] (OECD, 2010). For example, the price elasticity of demand for agricultural inputs is relatively inelastic meaning that low level taxes on pesticides or fertilisers in OECD countries have in many cases led to raising revenue but little change in behaviour. Using computer models, pollution charges could be directly proportional to the amount of diffuse pollution generated and set at a level where the marginal cost of reducing pollution is equal to the marginal benefit of emitting it.

In line with the polluter pays principle, water pollution taxes and charges should account for the following costs: *i)* direct costs (e.g. infrastructure, clean-up, wastewater treatment and drinking water treatment costs, and administrative, monitoring and data analysis costs); *ii)* external costs (e.g. negative environmental externalities such as reduced freshwater biodiversity and ecosystem functioning); and *iii)* opportunity costs associated with exclusion of other potential users in areas where water quality is unsuitable for use. In principle, revenue raised from such a regime should feed into the general budget of government and be applied to the highest priority public use. Some requisites for the design of water pollution charges and taxes are summarised in Table 4.5.

Table 4.5. **Some requisites for water pollution charges and taxes**

Requisite	Explanation
Clear objectives	The objectives of a pollution charge (i.e. a charge based on the quantity of pollutants that are discharged into the environment) or pollution tax (i.e. product charges on inputs that are believed to have environmentally harmful effects) should be clearly stated about what it aims to achieve: protecting the environment, raising awareness, re-balancing competitiveness across users, and/or raising revenue. The goal could be to ensure that water resources are used in a manner that maximises the net benefits, and that water of sufficient quality is available over time for its highest value use (economic, environmental and social). Obligations related to minimum water quality standards should be unambiguous.
Incentives to polluters	Water pollution charges (and taxes) should be linked to the quantity and toxicity of pollution discharged (or products used) to send a clear signal to users about the importance of water pollution reduction.
Reflection of environmental and opportunity costs, in line with the polluter pays principle	The level of the pollution charge or tax should reflect as best as possible the environmental and opportunity costs so that polluters get an accurate market signal about the costs of pollution. There is an economic rationale to differentiate the level of the charge/tax depending on the volume and toxicity of the discharge and the sensitivity of the local environment to pollution. Proxies can be used to estimate the negative externalities and opportunity costs associated with water pollution so that they can be reflected. Natural capital accounting may be a useful tool to assist with calculating appropriate charge or tax levels. Pollution charges and taxes should be indexed to inflation.
Equity	Differences in pollution charges or taxes should be based only on pollution characteristics (volume, toxicity, location, time) and the likely environmental and opportunity costs, rather than on the economic activity (e.g. agriculture versus industry) or the specific activity of pollution (irrigation versus industrial use).
Provisions for re-allocation of pollution allowances/ permits	Allowing polluters to trade pollution allowances both short term (within a season) and long term (for the duration of a permit) can improve efficiency in allocating abatement reduction effort.

Source: Adapted from Ambec, S. et al. (2016).

Water quality modelling also enables the ability to assign pollution allowances (or permits) much like regulation of point source pollution. Efficiency and equity impacts of allocating pollution in a performance-based regulatory setting differs across land uses and local contexts, and as such should be informed by economic analysis and stakeholder engagement.

Allocating diffuse pollution allowances within a cap then provides an opportunity for water quality trading (particularly in catchments approaching, or already at or above the cap). As discussed in Chapter 3, water quality markets can stimulate innovation, often achieve water quality targets at a lower social cost than traditional performance standards, taxes and payments/subsidies and eliminate the difficult task of setting pollution charge and tax levels. Some requisites for diffuse pollution allocation and for the design of water quality markets are summarised in Table 4.6.

Table 4.6. **Some requisites for diffuse pollution allocation and water quality markets**

Requisite	Explanation
Clear objectives and a stringent cap	Clearly state the objectives of a pollution cap, what it aims to achieve and why. The goal could be to ensure that freshwater ecosystems are restored/protected, and that water of sufficient quality is available over time for its highest value use (economic, environmental and social). Determine a cap at the catchment scale, informed by robust data and calibrated and validated modelling. Consider the assimilative capacity of water bodies, and the level of water quality required to maintain ecosystem functioning when setting a cap. Account for urban, industrial and rural sources of pollution (diffuse and point source) within a cap.
Identify allocation approaches	Identify allocation approaches (e.g. grandparent, catchment average, auction, natural capital - see Table 3.1). Consider including differences in the land resource as an integral part of sustainable water policy (see sub-section below). Foster stakeholder agreement on the principles of equity to use.
Assess the efficiency and equity implications	Estimate the catchment revenue impacts and benefits of each allocation approach to assess the relative efficiency. Evaluate the distributional impacts and opportunity costs of each approach for each land use. Account for abatement potential of the different land uses.

Table 4.6. **Some requisites for diffuse pollution allocation and water quality markets** *(cont.)*

Allocate pollution allowances	Select an allocation approach that best achieves efficiency and equity requirements. Allocate pollution allowances within the cap to all polluters in the catchment. Obligations related to complying with pollution allowances and monitoring and reporting should be unambiguous.
Compensate (if necessary)	Reflect the polluter pays principle where possible. Identify compensation mechanisms, if necessary, for those who face the highest costs or have the fewest mitigation options. Considerations for compensation should account for the abatement potential of different land uses, the ability to pay or farmer income, and lost opportunity costs.
Provisions for water quality trading	Allow polluters to trade pollution allowances both short term (within a season) and long term (for the duration of a permit) within a catchment to improve efficiency in allocating abatement reduction effort. To avoid potentially negative impacts of trade arising from changing the location of pollution, pollution allowances and trading arrangements must be consistent with the pollution cap. State clear rules from government to facilitate transactions. Provide for voluntary forfeiture of un-used pollution allowances. Keep transaction costs as low as possible relative to the anticipated nutrient prices and improvements in water quality. This requires limiting trading costs to administrative costs that are unavoidable and also limiting third party interference in individual transactions. Foster stakeholder engagement to create buy-in to the concept of trading.
Enable synergies with other policy sectors	Enable synergies between water quality and climate change mitigation and adaptation policies to fully benefit from complementarities and to minimise the risk of conflicts. For example, allowing stacking of nutrient credits with carbon credits in an emission trading scheme can further encourage innovation and co-benefits that reduce greenhouse gases and nitrogen pollution of water bodies at the same time.

Include differences in the land resource as an integral part of water policy

In setting policy for tackling declining water quality associated with diffuse pollution, policy makers need to consider the implications of water quality policy for economic growth and land use options into the future. The natural capital approach illustrated in Chapter 3 provides an alternative approach to allocating diffuse pollution limits based on current land use activities (such as grandparenting or sector average approaches), which have the potential to reward existing polluters and constrain future growth opportunities.

By recognising that land, like water, is a finite and critical resource, and that land differs in its productive potential and capability to filter and retain water and nutrients for plant growth, the policy driver can be changed from a resources efficiency use to one that recognises the necessity to add greater flexibility to landscapes that have little natural capital and/or lack versatility in either land use options and/or pollution mitigation strategies. In the long-term greater nitrogen reductions, water retention, productivity and therefore economic growth can be achieved by encouraging intensive land use activities on highly versatile soils, while phasing out intensive land use on poor quality soils. The higher the pollution reduction target, the more difficult it is to achieve improvements in water quality if intensive land use activities continue on lower quality soils. In essence, linking pollution allocation to soil characteristics will encourage over time a better match between inherent capability and use. In the short term, the approach can be costly if large investments in intensive farming activities have been made on poor quality soils. Pathways forward are required to manage the challenges associated with the transition to such a policy, in a similar way to which stranded assets in the climate sector need to be managed. Water quality trading can assist in making this transition.

The concept of adding ecological boundaries (e.g. a cap nitrogen losses to limit the impact on receiving environments), within which land use must operate, moves the analysis from managing land to managing a landscape connected to water. The ability to include ecological boundaries within which resources should be managed will be a feature and capability that analytical farm system frameworks will require in the future to reach the full economic potential of natural resources and provide multiple ecosystem services for a range of desired outcomes beyond just economic growth and water quality.

A role for central government

Central government will play a critical role in the transition to more effective management of water quality. Attention is shifting to the control of diffuse pollution, which has proven more challenging to monitor and regulate. The attainment of the following recommendations may expedite success:

- **Overarching national policy guidance and a strong direction** on water quality improvements is required to send the right signals to local authorities, stakeholders and investors. Distribute responsibility to achieve minimum water quality standards to local government and communities, which each have unique water quality issues, desired outcomes and capacities to respond.
- National policy guidance should be backed up by **regulatory frameworks and enforced minimum water quality standards** for setting the benchmark for better performance, and initiating innovations and investments in improving water quality. For example, minimum standards provide a benchmark, over and above which economic instruments can be used for water quality trading or payment for ecosystem services. Placing harmful chemicals on a watch list can encourage the innovation of more environmentally-friendly products. The amount of investment needed to meet new regulations should be considered when minimum water quality standards are developed. Without suitable funding, regulations cannot be met and their practical usefulness is limited.
- **Creating a space for stakeholder and community engagement** is necessary to manage perceived and actual risks, and reach solutions in partnership. Box 4.3 outlines some requisites for successful stakeholder engagement. Government transparency, accessibility of government services and information, and the responsiveness of government to new ideas, demands and needs are considered as the three building blocks to support an improved evidence base for policy making, strengthened integrity, lower corruption and higher trust in government (OECD, 2005).
- **Giving notice of policy changes and providing multiple options for implementation** of minimum standards is necessary to pave a way forward and reduce objections from stakeholders.
- **Providing government seed funding and allowing space for experimentation** (by relaxing regulations in such circumstances and distributing responsibility to local governments) can stimulate the diffusion of innovative technical and policy approaches that minimise the cost of water quality management. Examples may include pilots for wastewater reuse, water quality fit for purpose, decentralised systems, new approaches to manage and reduce diffuse pollution (e.g. nitrogen inhibitors, new cultivars, precision agriculture, constructed wetlands), and resource recovery from wastewater (i.e. energy and nutrients).

Box 4.3. Requisites for stakeholder engagement on water quality

Stakeholder engagement is necessary to achieve common objectives on water quality management. It identifies stakeholder preferences and desired outcomes, provides a constructive means for collective decision-making about sharing the risks, costs and benefits, and encourages buy-in and compliance with implemented policies. Stakeholder engagement is also required for policy integration, harmonisation, and governance to build synergies and generate co-benefits across sectors and public–private partnerships.

The success of collaborative approaches depends on:

- A national process/framework to ensure the appointment of collaborative groups reflect a balanced range of the community's interests, values and investments. Principles to guide stakeholder governance frameworks are provided in the table below.
- Having sound and transparent processes that encourage stakeholders to freely supply their knowledge and opinions, and encourages them to negotiate honestly. Some stakeholders may withhold important information and negotiations may not reach sufficient agreement, irrespective of the soundness of the facilitation process. Regardless of the quality of the collaborative process, some or all stakeholders may still be critical of the final policy design.
- Providing stakeholders with a clear understanding of the policy design task and process, and with knowledge of the requisite design tools and skills. How policy design proceeds, and the tools used in that process, then becomes the shared territory of the collaborative parties and the decision-making authorities.
- Honest "knowledge brokers" or creating a space for the brokering, where there is competing science and/or entrenched views that block action and policy development.
- Sufficient funding to compensate employers whose staff may be chosen to represent community interests. Funding should also be supplied to secure independent experts for technical scientific investigation to inform decision-making.
- Tools to evaluate, track and report on the progress of collaborative governance in line with the OECD principles.

OECD Principles on stakeholder engagement in water governance

Principle	Description
Inclusiveness and equity	Map all stakeholders who have a stake in the outcome or that are likely to be affected, as well as their responsibility, core motivations and interactions
Clarity of goals, transparency and accountability	Define the ultimate line of decision making, the objectives of stakeholder engagement and the expected use of inputs
Capacity and information	Allocate proper financial and human resources and share needed information for result-oriented stakeholder engagement
Efficiency and effectiveness	Regularly assess the process and outcomes of stakeholder engagement to learn, adjust and improve accordingly
Institutionalisation, structuring and integration	Embed engagement processes in clear legal and policy frameworks, organisational structures/principles and responsible authorities
Adaptiveness	Customise the type and level of engagement as needed and keep the process flexible to changing circumstances.

Sources: Kaine and Boyce (2015); OECD (2015b).

Note

1. Fertiliser taxes can cause an additional burden on horticulture production while making livestock production more profitable. They may also provide unintended incentives to increase livestock levels, leading to greater manure production through more intensive protein feeding, larger acreages devoted to nitrogen-fixing plants and reorganisation of crops in favour of those with less nitrogen consumption, but not necessarily less nitrogen surplus (OECD, 2010).

References

Ambec, S. et al. (2016), "Review on international best practice on charges for water management", Toulouse School of Economics, OECD Background Paper (unpublished).

Anastasiadis, S. et al. (2013), Does complex hydrology require complex water quality policy?, *Australian Journal of Agricultural and Resource Economics*, Vol. 58, pp. 130–145.

Brears, R. (2016), Korea's smart grids for water and energy, http://markandfocus.com/2016/02/25/koreas-smart-grids-for-water-and-energy/ [accessed 31/03/2016].

Defra (2011), Press release: £110 million revamp for England's rivers, Department for Environment, Food & Rural Affairs (Defra), UK.

Fishmana, Y., Beckerb, N. and M. Shechterc (2012), The Polluter Pays Principle as a policy tool in an externality model for nitrogen fertilizer pollution, *Water Policy*, Vol. 14, pp. 470–489.

Grey, D. et al. (2013), Water security in one blue planet: Twenty-first century policy challenges for science. Phil. Trans. R. Soc. A 371.

IBM (2010), (2012), Creek Watch iPhone App, https://www.ibm.com/blogs/research/2012/03/creek-watch-iphone-app-goes-social/ (accessed 23/03/2016).

Kaine, G and W. Boyce (2015), Designing Policy to Change the use of Natural Resources, Policy Brief No.11 (ISSN: 2357-1713), Landcare Research, New Zealand.

OECD (2016a), *Water Governance in Cities, OECD Studies on Water*, OECD Publishing, Paris. http://dx.doi.org/10.1787/9789264251090-en.

OECD (2016b), *Recommendation of the Council on Water*, 13 December 2016 - C(2016)174/REV1/FINAL.

OECD (2015a), *OECD Water Governance Principles*, www.oecd.org/gov/regional-policy/OECD-Principles-on-Water-Governance-brochure.pdf.

OECD (2015b), *Fostering Green Growth in Agriculture: The Role of Training, Advisory Services and Extension Initiatives, OECD Green Growth Studies*, OECD Publishing, Paris, http://dx.doi.org/10.1787/9789264232198-en.

OECD (2013), *Water Security for Better Lives*, OECD Studies on Water, OECD Publishing, Paris, http://dx.doi.org/10.1787/9789264202405-en.

OECD (2012a), *A Framework for Financing Water Resources Management*, OECD Studies on Water, OECD Publishing, Paris, http://dx.doi.org/10.1787/9789264179820-en.

OECD (2012b), *Water Quality and Agriculture: Meeting the Policy Challenge*, OECD Studies on Water, OECD Publishing, Paris, http://dx.doi.org/10.1787/9789264168060-en.

OECD (2010), *Taxation, innovation and the environment*, OECD Publishing, Paris, http://dx.doi.org/10.1787/9789264087637-en.

OECD (2009), *Managing Water for All – An OECD Perspective on Water Pricing and Financing*, OECD Publishing, Paris, http://dx.doi.org/10.1787/9789264059498-en.

OECD (2005), *Modernising Government: The way forward*. OECD Publishing, Paris, http://dx.doi.org/10.1787/9789264010505-en.

OECD (1974), *Recommendation of the Council on the Implementation of the Polluter-Pays Principle*, 14 November 1974 - C(74)223 - C/M(74)26.

OECD (1972), *Recommendation of the Council on Guiding Principles concerning International Economic Aspects of Environmental Policies*, 26 May 1972 - C(72)128 - C/M(72)15.

Shortle, J. and R.D. Horan (2017), Nutrient pollution: A wicked challenge for economic instruments, *Water Economics and Policy*, Vol. 3, No. 2.

USGS (2015), Water-Quality Data, http://water.usgs.gov/owq/data.html [accessed 31/03/2016].

ORGANISATION FOR ECONOMIC CO-OPERATION AND DEVELOPMENT

The OECD is a unique forum where governments work together to address the economic, social and environmental challenges of globalisation. The OECD is also at the forefront of efforts to understand and to help governments respond to new developments and concerns, such as corporate governance, the information economy and the challenges of an ageing population. The Organisation provides a setting where governments can compare policy experiences, seek answers to common problems, identify good practice and work to co-ordinate domestic and international policies.

The OECD member countries are: Australia, Austria, Belgium, Canada, Chile, the Czech Republic, Denmark, Estonia, Finland, France, Germany, Greece, Hungary, Iceland, Ireland, Israel, Italy, Japan, Korea, Latvia, Luxembourg, Mexico, the Netherlands, New Zealand, Norway, Poland, Portugal, the Slovak Republic, Slovenia, Spain, Sweden, Switzerland, Turkey, the United Kingdom and the United States. The European Union takes part in the work of the OECD.

OECD Publishing disseminates widely the results of the Organisation's statistics gathering and research on economic, social and environmental issues, as well as the conventions, guidelines and standards agreed by its members.

OECD PUBLISHING, 2, rue André-Pascal, 75775 PARIS CEDEX 16
(42 2017 07 1 P) ISBN 978-92-64-26905-7 – 2017

www.ingramcontent.com/pod-product-compliance
Lightning Source LLC
Chambersburg PA
CBHW061221240926
51846CB00006B/1309
* 9 7 8 9 2 6 4 2 6 9 0 5 7 *